STEM for All Ages

How Science, Technology, Engineering & Math drive progress

By

Seán G. Dwyer

AuthorHouse™ LLC
1663 Liberty Drive
Bloomington, IN 47403
www.authorhouse.com
Phone: 1-800-839-8640

Published by AuthorHouse   02/04/2014

ISBN: 978-1-4918-5880-6 (sc)
ISBN: 978-1-4918-5881-3 (e)

Library of Congress Control Number: 2014902234

Table of Contents

Preface

The history in this book will surprise many readers. Do you know the 1st names of the brothers who invented the first successful manned aircraft? If you answered "Orville and Wilbur", you would be in good company, but wrong. "Joseph and Etienne" would be more correct. Ben Franklin wrote about them, and their type of aircraft is still in wide use today. However, 150 years before that, the Celebi brothers survived the first inter-continental flight and first manned rocket flight. Surprised? If so, that is good, you remember surprises, but the details would help.

Roughly every 500 years the introduction of new technology completely disrupts society. Everything changes, borders, laws, currency, and the way people work. It happened again when we transitioned from the Industrial Age to the Digital Age with the introduction of the microchip.

With the goal of providing a more interesting learning experience, the focus in this book is not just on STEM (Science, Technology, Engineering, Math). Also included is analysis of Cause & Effect. So it is not just about HOW things work, it is also about WHY they happened that way, and the consequences. For example, how come Native Americans never invented the wheel? How could a Viking farm in Greenland support 150 cows? While the topics might appear unrelated to each other, they are united by the theme of Cause & Effect.

Most of the chapters originated from talks or classes presented by the author to audiences ranging from grade school to post-graduate university classes, engineers, aviators, business groups, and retirees. Many of the STEM chapters were developed for the youth programs of Experimental Aircraft Association Chapter 838 in Racine, WI.

Because aviation involves many fields of science, it is a particularly interesting way to show how STEM is a continuum of mutually supporting elements. Science would be just a hobby without engineering. Engineers use science as a template when they design or fabricate airplanes, bridges, whatever. Technology is that 'whatever'. It is the output when science is combined with engineering. Finally, math is just a language that scientists, engineers, and technologists use to communicate. When unleashed, STEM is enabling. But what leashes it? Knowing that is important. The concept of paradigms and paradigm paralysis is discussed.

Not only does Archimedes' Law explain why balloons go up and submarines go down and come up again, but planes would not be controllable without his Lever Principle. Chemistry's Ideal Gas Laws explain how the first aircraft flew and why planes need longer runways on hot days. Newton's Laws of Motion explain the lift of airfoil shaped airplane wings, but airfoils need Bernouilli's Principle and that would not work in air without the Coanda Effect. All are easier to demonstrate than explain and this book strives to do both.

Just as the Industrial Age was enabled by the automation of factories by steam engines, the Digital Age was created by the microchip. Microchips provide the basis for three of the four economies in today's world, which Kenichi Ohmae calls *The Invisible Continent.* They include the *Visible Economy* (i.e. buy a book in a shop), *Cyber Economy* (buy it from Amazon.com), *Borderless Economy* (where is Amazon located?), and *High Multiples Economy* (why so many billionaires). Today, most people live and work in the *Invisible Continent,* whether they know it or not. Understanding its rules – or lack of them – is valuable.

Young people today will experience a faster rate of progress than those of us who grew up in the last days of the Industrial Age. My Grandmother was born in 1888 and died in 1974. Already 15 years old when the Wright brothers made their first flight, she was 81 when man walked on the Moon. What progress in one lifetime! What will today's 15 year olds experience?

The chapter on mega-history explains how one technology (stirrups) enabled the Feudal Era, which was then eliminated by another technology (cannon guns). Cannon guns fostered the growth of nation states and many forms of government, which are rapidly becoming obsolete with the advent of the Digital Age. Microchips will have as far-reaching effects on society as cannon guns had.

The chapter on weather explains why we have deserts, rain forests, Trade Winds, and why you can grow strawberries as far north as Norway. Another chapter addresses the global warming controversy. We are between Ice Ages, so we are either warming or cooling. Which would you prefer? True or not, it is all Cause & Effect. To meet the challenges of the Digital Age, people young and old would do well to understand both the HOW and the WHY.

If you don't recall something, have you really learned it? You certainly can't use it. Since I was a young boy I have always been interested in science, but it was the application of science that made it both memorable and understandable. Sometimes that took the form of trivia, which is why I have inserted chapters of Q&A throughout. It is my hope that this book will make STEM both understandable and easier to remember, and something to be pursued without fear or prejudice.

Chapter 1   # STEM + Cause & Effect

As mentioned in the Preface, my Grandmother was 15 when the Wright brothers first flew, and 81 when man walked on the Moon. Not only was that rate of progress simply amazing, it was vastly faster than in the millennium before. The Industrial Age was less than 100 years old at the time of her birth, and progress in her lifetime was not just technological. Society, governments, borders, and ways of life also changed. When she was born in Ireland in 1888, Ireland was still part of the United Kingdom, Victoria was Queen, and the Sun never set on the British Empire.

Six other empires existed at the time. One was ruled by Victoria's grandson, the German Kaiser. In another, the Russian Tsar was married to her granddaughter. Four other emperors were the Austro-Hungarian Emperor, the Ottoman Sultan, and the Emperors of China and Japan. During my Grandmother's life all seven empires were relegated to garbage bins of history and more than a hundred new countries emerged. Just as much change is ahead of us.

When I showed my grandchildren my Father's stamp album and my own childhood album, it was as though we were looking at stamps from different planets. Countries like Aden, Indochina, Yugoslavia, Czechoslovakia, Southern Rhodesia, USSR, UAR, Gold Coast, and many more all came and went during the periods covered by the albums.

Science, Technology, Engineering, & Math

In contrast to the rate of progress experienced in my Grandmother's lifetime, Marco Polo saw rockets in 1265 AD. Yet, hundreds of years were to pass before men used them to fly. Why so long? The answer is that STEM was fragmented. They had scientists and mathematicians, but not enough engineers, and that meant that technology was slow in coming. This is a little surprising, as the Roman Empire clearly had engineers. You just have to look at the Coliseum to appreciate this. Why were there so few engineers in the Dark Ages that followed the fall of the Roman Empire. Don't be too quick to conclude that the Dark Ages were caused solely by the decline of the Roman Empire. Weather, or more specifically, climate, also played a role. That will be discussed later.

Progress in the last 250 years was catalyzed by knowing how to use laws of chemistry and physics in what I call the STEM continuum, and having the resources to do so. If you blow up a balloon and release it, it will fly around the room in a perfect demonstration of **science**, specifically Newton's 3rd Law:

Every action has an equal and opposite reaction.

Air coming out of the balloon is action, and the balloon flitting around the room is the reaction.

Science is the development or discovery of such laws. Science tells you that if you blow enough air back, an airplane will go forward. If its wings push enough air down, an airplane will go up. However, that is just enough information to get you killed if you do not have control. That is where engineering comes into play.

Engineering includes design and fabrication based on established rules of science. Engineering is the application of science and is what makes science useful. An example is the design and manufacture of an airplane's wings and ailerons to provide lift and control, all based on the science of the Principle of the Lever combined with Newton's 3rd Law.

Technology is the output when you combine science and engineering, e.g. an airplane, a bridge.

Math is the language whereby science, technology, and engineering communicate with each other. Math includes equations that are centuries old and computer programs that are still being developed. Examples include $\pi\,r^2$ which is the area of a circle, **Arm x Force = Moment** the equation for the Principle of the Lever, and **PV = nRT** which reflects the Ideal Gas Law. This brings me to Cause & Effect.

Cause & Effect

While politicians harangue about the off-shoring of jobs, the truth is that far more jobs are lost due to technological obsolescence. The next time you see an ATM ask it how the summer is going. Do not be offended when it doesn't answer. It is a machine, and the jobs of the bank tellers it replaced are not coming back. Similarly, getting a job as a travel agent or manual labor in a car factory would not be good career choices, although being a programmer of the machines that build cars would be. Take politicians with a grain of salt.

Why did the Wright brothers put the elevator (Fig. 1) in front and their propellers and rudder at the back of the 1903 Wright Flyer, and what was the effect of doing so? Why was the elevator of their 1911 Wright Flyer aft with the rudder, where elevators have been for most airplanes ever since?

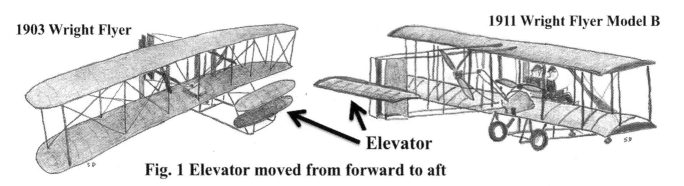

1903 Wright Flyer

1911 Wright Flyer Model B

Elevator

Fig. 1 Elevator moved from forward to aft

Analyzing such Cause & Effect issues provides an opportunity to create better designs and overcome paradigm paralysis. To illustrate this point, the Wright brothers had never seen a propeller or rudder anywhere but at the back of a boat, so their paradigm (i.e. a thought pattern or rule that one follows without knowing it) was to put the rudder and propellers where they were in the 1903 Wright Flyer. But they had never seen an elevator, and called theirs a "horizontal rudder". It was logical (per the Lever Principle) to put the elevator in front to balance the "vertical rudder" at the back, or perhaps they wanted a shock absorber in front. Either explanation could have been the "cause". What was the "effect"?

Having the elevator in front was like trying to push a rope. The faster the plane went, the more distortion would occur. Making the elevator stronger would have added weight. Bottom line, one effect of having the elevator in front was a severe limitation to airspeed. Understanding this limitation and its cause offered its own solution. Since pushing up on one arm of a lever is the same as pushing down on the other arm, why not put the elevator at the back?

In the 1911 Wright Flyer Model B the elevator was positioned aft of the rudder. While a good solution for the time, it also brought in a cost (to be covered later in the chapter on airplane physics) that was not seriously challenged for over 70 years because of paradigm paralysis. When it was, the result was lighter, faster, and more fuel-efficient airplanes. This is a good example of why Cause & Effect analysis is so valuable.

Contrary to popular belief, the Wright brothers were not even close to being the first to fly. Two other brothers had invented the first aircraft more than a century earlier, and hundreds of people flew before the Wright brothers.

Knowingly and unknowingly, applications of science impact many aspects of our daily lives. For example, what is the connection between an open-top freezer in a grocery store and a hot air balloon? What is the connection between the Goodyear Blimp and a building code that requires a step up from the garage into a house?

In the case of the tub freezer and hot air balloon, the connection is Charles' Law, which states:

At constant pressure an ideal gas expands in direct proportion to its absolute temperature

Just as the tub freezer is open at the top, a hot air balloon is open at the bottom, so the air in both is at the pressure of the surrounding atmosphere. The hot air in the balloon expands leaving fewer molecules of hot air trapped inside the air bag, so the balloon is lighter than the surrounding air. The opposite is true with the tub freezer. The cold air in it contracts, resulting in more molecules per unit volume, so the cold air is heavier. Thus a lid is not needed to keep the cold air inside the freezer.

In the case of the Goodyear Blimp and the building code, the connection is another of the Ideal Gas Laws. Avogadro's Law states:

Equal volumes of different gases contain the same number of
molecules, as long as the temperature and pressure are the same

The Goodyear Blimp is filled with Helium which has a molecular weight of 4. This is vastly lighter than air, which has an average molecular weight of 28.6. So the blimp can fly. On the other hand, gasoline vapors are much heavier than air, e.g. octane has a molecular weight of 114. Garages are essentially gasoline storage facilities and should be treated as such. Does the gasoline in the car, the lawn mower, or the can on the shelf ever drip out? If it does, you do not want those explosive (heavy) vapors entering the house. That is the role of the step up from the garage.

Are these examples useless scientific trivia? Knowing the second one might cause you to hesitate before storing gasoline in a basement, particularly one with a sump pump. Sooner or later, the heavier vapors from a spill would make their way to the lowest point in the basement, where they would wait for a spark the next time the sump pump activates. More importantly, knowing the laws of science and how to apply them provides solutions to many problems.

Chapter 2   **The First Aviators**

21 November, 1783

"Condemned criminals?" shrieked the foppish Frenchman at Etienne Montgolfier, as he flung his perfumed handkerchief at the ceiling. Pilatre de Rozier was incensed when he heard that King Louis XVl had agreed to the "loan" of two condemned criminals for the first flight of the Montgolfier hot air balloon with people on board.

"Criminals don't deserve the honor of being the first aeronauts!" asserted de Rozier. *"I'll do it!"*

Meanwhile, Etienne's brother, Joseph Montgolfier, watched as the handkerchief spread out and floated down.

"What if we attached cords to the four corners of a bed sheet. Could we use it to escape from the balloon if it catches fire?" he wondered. Ever the dreamer, Joseph had heard stories about a Turk named Celebi, who floated safely down "on the wings of eagles" after blasting himself into the air with rockets. Celebi survived that flight in 1633. Joseph did not believe the 'wings of eagles' bit, but we will get to that story later.

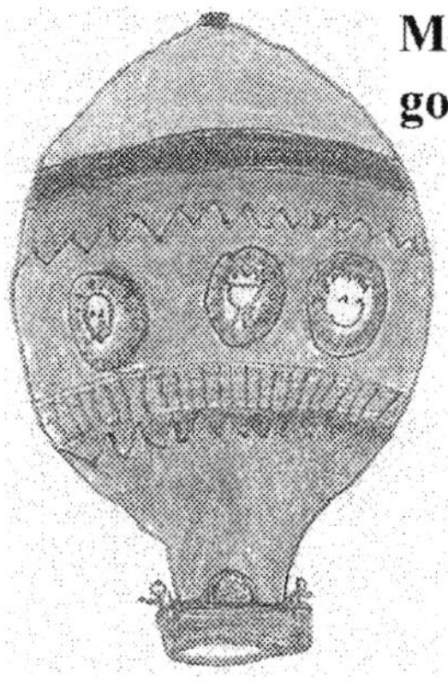

Pilatre de Rozier, a teacher of chemistry and physics, had a right to be angry, as he had already gone up in the balloon while it was still tied to the ground. However, being aloft in a tethered balloon was no more 'aeronautical' than climbing a ladder. He was fairly confident that flight itself would not be fatal, as long as they did not go up too high, although coming down safely might be a problem. Two months earlier, in the courtyard of the Palace of Versailles with the King in attendance, he had assisted in an untethered flight in which the passengers were a duck, a cock, and a sheep.

Montgolfier Hot Air Balloon got its lift per Charles' Law

"At constant pressure the volume of a gas is directly proportional to its absolute temperature"

Fig. 1

The reason for selecting those particular animals as test pilots was simple.

Nobody knew if going up in the air would be survivable for a creature – like man – which was not designed to fly. They knew that the duck could fly, so it should survive the flight, whatever about the landing. The cock, like any chicken, was designed to be able to fly, but rarely did so. So his chances were good. The sheep, on the other hand was clearly not designed for flight. If the sheep survived, then man's prospects were good. All three test pilots survived, although stories about flying sheep that spread among peasant farmers may have been the cause of problems for later balloonists.

Pilatre de Rozier prevailed. The decision was communicated to King Louis XVI that de Rozier would be accompanied by Marquis d'Arlandes on the first untethered, manned flight. The flight (Fig.1) took place from a garden in the Bois de Boulogne in Paris. The King was in attendance, as were many nobles, scientists, and commoners, half the population of Paris in fact. Ben Franklin watched the proceedings from his hotel room window. He was in Paris in 1783 for a reason that was very important for America. He was negotiating the Treaty of Paris, which ended the American War of Independence. Incidentally, many of the people there that day would also be cheering when they saw the King's head being chopped off less than 10 years later.

The 25 minute flight attained an altitude of 3,000 feet and traveled over 5 miles to the outskirts of Paris. For the first time in history, man had flown, landed safely, and could do it again in the same aircraft. Its ability to fly was explained by **Charles' Law**, which was then emerging as a law of chemistry, and by **Archimedes' Law**, a 2,000 year old law of physics.

Another thing that might surprise you, was that the French scientist, Jacques Charles who wrote the law, was really upset by the success of the Montgolfier balloon. You might think he would be pleased. The reason for his displeasure was that he was leading another team in the competition to be the first aviators in history.

Charles' Hydrogen Balloon relied on Avogadro's Law

"Equal violumes of ideal gases at the same temperature & pressure have the same number of molecules"

Fig. 2

Jacques Charles and his team were already filling a hydrogen balloon (Fig. 2), which would have much higher performance, but did not make its first manned flight until December 1, 1783, less than two weeks after the Montgolfier balloon. The lift of Charles' balloon was explained by **Avogadro's Law**, another one of chemistry's **Ideal Gas Laws**. The sciences that allowed balloons to fly will be covered again in Chapter 3.

Charles' balloon was much smaller than the Montgolfier balloon, and its first flight flew 30 miles over two hours with Jacques Charles and Nicholas Robert on board. After descending by releasing some of the gas, Nicholas Robert stepped out of the basket. The reduced weight in the basket caused the balloon to rise again. Charles flew solo and the lighter aircraft ascended to 9,000 feet.

The Golden Age of Balloons had begun, and many hobbyists in France began experimenting with both hot-air and gas balloons. Needless to say, they tended to be wealthier people. One balloonist – the French called them aeronauts – landed among a group of peasant farmers working in a field. These peasants had heard stories about flying sheep, and naturally assumed that witchcraft was involved. They tore the balloon apart and nearly killed the aeronauts on board.

Word went out in the flying community to have a 'Plan B' in the event of a descent among terrified farmers. In case you ever wondered why a glass of Champagne is always served after a balloon lands, now you know. It was Plan B. When you scare the daylights out of a groups of peasants armed with hay pikes and scythes, immediately offer them a glass of Champagne. It is still done to this day, although the origin of the practice is lost in the mists of history. Of course today it is more "Thanks for letting us land unexpectedly in your field" rather than "Really, we are not witches".

The Golden Age of Balloons

Pilatre de Rozier was a scientist and a test pilot for both the first tethered flight and first free flight of the Mongolfier balloon. His experiments with balloons continued, but he was killed in 1785 in an attempt to fly from France to England across the English Channel. This balloon combined a hydrogen bag with a hot air bag fueled by burning straw and wool. That is a deadly combination, because any leaked hydrogen coming into contact with the combustion source would go boom. This combination of a lighter-than-air gas and hot air balloon is known as a 'Roziere'. Of course, it would be much safer if the gas in question was non-flammable helium rather than hydrogen. The discovery of helium would not occur for about a century.

Surprisingly, more than 200 years later balloonists striving to be the first to circumnavigate Earth used Rozieres. However, rather than flammable hydrogen, they all used the inert gas helium in a sealed bag inside a larger bag of heated air. The helium would heat up and expand during the day and cool off and contract at night, so the gas bags were never filled to bursting point. Understanding the relationships between pressure, volume, and temperature was key for them. We will get back to that in the chapter titled "The M in STEM".

The first balloon to fly all the way around the Earth non-stop was the Breitling Orbiter 3 in March 1999, 216 years after the first manned balloon flight. Bertrand Piccard and Brian Jones took 19 days, 1 hour and 49 minutes for the flight. Three years later, Steve Fossett did it solo.

Fig. 3 Blanchard & Jeffries 1785

Shortly before de Rozier's death another team successfully crossed the English Channel by balloon (Fig. 3), although this was from the English side to the French side. The pilots were a Frenchman Jean-Pierre Blanchard and an American, Dr. John Jeffries. Their story would make a good movie. Although Jeffries financed the flight, Blanchard wanted to avoid having to share the glory. As they made ready to lift off, they found that the H_2 balloon did not have enough lift. The suspicious Jeffries searched Blanchard and found that he was wearing a leather belt with lead weights hanging from it. Once these were removed, they were able to depart. However, it was a close thing and the balloon nearly hit the water several times. They threw everything that was not essential overboard. This included the silk covered oars and the hand operated, six bladed propeller, which Blanchard called a "moulinet" (little windmill). The final items to go overboard were their coats and pants. They arrived near Calais in their underwear.

What they were trying to create with their oars and the propeller was a "dirigible", a directionally guidable balloon. However, the spherical shape and large size of the balloon made directional control very hard to achieve, and they just had to go where the wind blew them.

In that 1785 flight Blanchard and Jeffries carried a letter from Ben Franklin's estranged son William, who lived in London, to Ben and Temple Franklin who were in France. William had been the last Colonial Governor of New Jersey and remained a loyalist throughout the American War of Independence, eventually accepting exile from the new USA to live in London. William, who was born out of wedlock, was raised by Ben and was the son who helped in his famous kite experiment in 1752. Temple was William's out-of-wedlock son and was raised by Ben. Marriage as we know it today may not have been a high priority among the Franklins, but they took responsibility for their off-spring.

In 1793 Blanchard made the first balloon flight in America, carrying a letter from President George Washington to wherever he landed in New Jersey. The letter provided an introduction of Blanchard to whomever was present where he landed. Presumably, the fact that it was signed by the very popular of President of the United States would convince the reader that Blanchard was not a witch, a fear that had already resulted in casualties in France.

This letter cemented Jean-Pierre Blanchard's position as the 'father of air mail'. It also meant that George Washington was the first president to send an airmail letter, whereas Founding Father Ben Franklin had been the first person to receive one. Others present that day in 1793 included two future presidents, Vice President John Adams and Secretary of state Thomas Jefferson.

Union Army Balloon Corps 1863

Americans balloonists of note included Thaddeus Lowe (1832-1913) who was planning a trans-Atlantic flight when the Civil War broke out. He offered his services to Abe Lincoln for artillery spotting and in 1861 he ran a demonstration from the Armory, which was across the street from the White House. Lowe formed the Union Army's Balloon Corps (Fig. 4), but

Fig. 4 Union Army Balloon Corps

went home with his balloon in 1863 after figuring out that he was "the most shot at man in the Union army." If he could see them, they could see him.

The Union Army's Balloon Corps was not the first use of aircraft in war, as the first air force in the world was the French Company of Aeronauts (Compagnie d'Aerostiers in French) was formed in 1794. Napoleon was not impressed with the results and abandoned the concept.

Aereon 1864

Solomon Andrews (1806-72) was a contemporary of Thaddeus Lowe and twice tried to get Lincoln interested in his very innovative balloon. Because it was capable of flying against the wind and directed flight, it would be more accurate to describe it as a "dirigible". Surprisingly, it did not have an engine of any kind. Its power was a combination of gravity and Archimedes' Law.

The first Aereon (Fig. 5) had three 80' long tubes attached side by side, with seven smaller bags of hydrogen in each tube. A basket suspended underneath held the pilot who sat on a seat that could move forward and back on rollers. By shifting the center of gravity in the basket, Andrews could vary the angle of attack of the balloon.

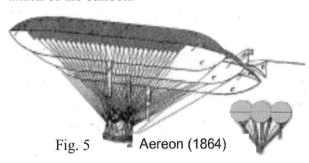

Fig. 5   Aereon (1864)

While a spherical balloon would climb vertically when ballast is dumped, a flat balloon raised at one end would move in the direction of the raised end as it ascended. Andrews described his method of propulsion as "gravitation" and varied buoyancy by dumping either gas or ballast. He would fly forward in a series of up and down glides similar to tacking in a sail boat, except that his balloon tacked vertically whereas a sail boat tacks from side to side.

To see how it worked all you need is a ball, a board, and a swimming pool or bathtub. Hold the ball against the floor of the pool and then release it. It will shoot straight up, as you would expect from Archimedes' Principle. Now try the same thing with the board, but before you release it, raise one end of the board higher than the other end. What you will observe is the board zooming off in the direction of the higher end as it rises through the water. Unlike the board in water, the Aereon could also be made to descend by releasing some of the gas.

Thus, the Aereon used the buoyancy promised by Archimedes' Law to ascend, and the angle-of-attack of the raised end plus air resistance from above to drive it in a forward direction. Once the Aereon had achieved its maximum altitude where buoyancy would be neutral, release of hydrogen from the gas bags would allow gravity to pull the aircraft down. Once again, wind resistance – but from below this time – would cause the Aereon to continue moving forward, even against the wind. Turns were achieved by using the rudder at the back, and 360 degree turns were demonstrated.

An additional design element was the longitudinal cavities formed above and below where two gas cylinders met. Since there were three cylinders in Aereon 1, there were two of these cavities above and below, and they acted like rails as the aircraft ascended or descended. The high pressure air forced into the cavity when ascending would travel the length of the cylinders and exit out the back, providing focused forward boost. This feature was dispensed with in Aereon 2 which had the shape of a flattened lemon, sharply pointed at both ends.

Whereas Aereon 1 descended by releasing gas, Aereon 2 had a complicated system of controlling altitude by using cords and pulleys to expand and contract the hydrogen gas bags, thereby changing how much air was being displaced. Keep in mind that, unlike a child's toy balloon which you blow up, the gas in Andrews' balloon was not under pressure.

Even a direct headwind would not prevent the Aereon from rising and advancing, unless the wind speed was faster than the forward motion produced by the sloping ascent of the Aereon. A newspaper account on Sept 4, 1864 described test flights by Solomon Andrews in a balloon containing 26,000 cubic feet of hydrogen and 256 lbs. of ballast at the start. Andrews made multiple ascents before finally stepping out after lashing the rudder to one side and pointing the bow up. According to the newspaper report: "She ascended at a tremendous rate of speed, computed at over 200 miles an hour at the beginning, in a spiral of say half a mile across, until lost to sight among the higher clouds".

If you find this hard to believe, read it yourself at the following link: http://www.nytimes.com/1865/06/11/news/local-intelligence-aerial-navigation-aercon-invented-dr-andrews-new-jersey.html?pagewanted=1

On July 5, 1864 Patent # 43,449 was issued by the US Patent Office for a balloon that was capable of directed flight and could even be flown against the wind. (See Appendix 1)
Solomon Andrews' two attempts to get Abraham Lincoln interested were blocked by the Whitehouse staff. Quite simply, they never let any of his letters reach Lincoln's in-box.

Andrews made enough widely reported controlled flights around New York that we can be sure that Aereon flew. But why did why did history ignore this engineless dirigible? I can think of three reasons.

The first was the assassination of President Lincoln on April 14, 1865. When Solomon Andrews made his flights in June of that year, the newspapers were still full of the stories about the conspirators, their trials, the reactions of foreign leaders, etc. Just as few people in Britain saw the first episode of *Dr. Who* on November 23, 1963, few paid any attention to stories about Solomon Andrews' amazing aircraft. (In case you missed the reference, November 23, 1963 was the day President Kennedy was assassinated)

The Aereon certainly fit the definition of a 'dirigible', a directionally controllable balloon. All other dirigibles had an engine of some kind, and the first one, Henri Giffard's Airship in 1852, had preceded the Aereon by more than a decade. Perhaps that is why history ignored it.

Alternatively, Aereon ll may have seemed too much like a perpetual motion machine because its forward motion was at least in part provided by compressing and expanding the hydrogen bag, and Aereon 1's directed travel required release of ballast and hydrogen, i.e. fuel was consumed. But that is a reality for all powered vehicles.

Solomon Andrews' ideas would resurface 100 years later (Aereon Hybrid Airship), and again 50 years after that (Aeroscraft). We will get to them in the chapter on dirigibles. It might surprise you to learn that the first revenue generating airliner was a dirigible, as was the first trans-Atlantic airliner.

Before we get to dirigibles, we should cover the sciences that enabled the first successful AND reproducible aircraft. You will see the reasons for the qualifier in the previous sentence in later chapters, "The First Airplanes" and "The Celebi Brothers Vs. Gravity". Basically, for a flight to be considered successful, you have to be able to walk away afterwards and do it again in a reasonable period of time. Many pioneers were carried away from their first and only flight.

Chapter 3

Science of the First Aviators

Ideal Gas Laws of Chemistry

I think it was the Ideal Gas laws that made me want to be a chemist. Not only do they fit together perfectly like pieces of a jigsaw puzzle, they changed the world. The three laws, Boyle's, Charles', and Avogadro's, combine to yield the Ideal Gas Law. These science laws helped pave the way to the Industrial Age.

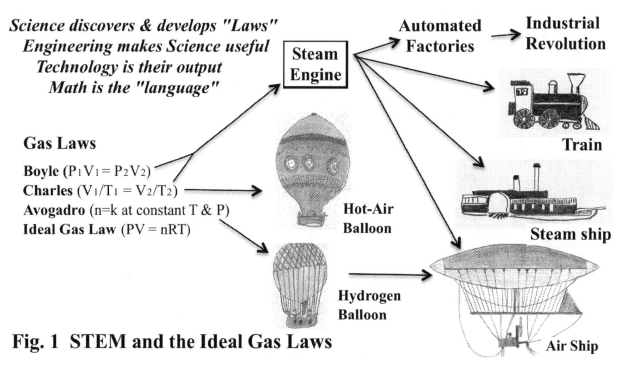

Science discovers & develops "Laws"
Engineering makes Science useful
Technology is their output
Math is the "language"

Gas Laws

Boyle ($P_1V_1 = P_2V_2$)
Charles ($V_1/T_1 = V_2/T_2$)
Avogadro (n=k at constant T & P)
Ideal Gas Law ($PV = nRT$)

Steam Engine

Automated Factories

Industrial Revolution

Train

Hot-Air Balloon

Steam ship

Hydrogen Balloon

Air Ship

Fig. 1 STEM and the Ideal Gas Laws

The collage in Fig.1 perfectly demonstrates the term STEM, which stands for Science, Technology, Engineering, and Math. The four laws on the left are SCIENCE. The machines and balloons are examples of TECHNOLOGY. The process of designing and fabricating them was ENGINEERING. Finally, MATH is the language that ties them all together, and is represented by the equations that reflect the laws. Bottom line, STEM changed the world with the gas laws of chemistry.

Boyle's Law

Irishman Robert Boyle (1627-1691) was known as "the Father of Chemistry". Although his law might seem obvious now, it was not at the time. He demonstrated via experiments that the pressure (P) of a fixed amount of gas multiplied by its volume (V) would be a constant as long as temperature remained the same. This is expressed mathematically as:

$$P \times V = k \text{ where k is a constant number.}$$

In other words, if you double pressure on a quantity of gas, its volume drops to half what it was before pressure increased, as long as temperature remains the same. Boyle's Law says:

The volume of an ideal gas is inversely proportional
to its pressure when temperature remains constant.

Another way to express Boyle's Law mathematically is $P_1V_1 = P_2V_2$, where subscript numbers represent a different set of pressure and volume for the same quantity of gas, but with both sets of readings being taken at the same temperature. However, anybody who has ever used a pump to inflate a tire knows that the pump will heat up, so it is not quite as simple as it first appears. This is where Charles' Law comes into play. Remember, I said that the Gas Laws were like a jigsaw puzzle where all the pieces fit together.

Charles' Law

Jacques Charles (1746–1823), a French scientist, led the team that came second in the race to become the first aeronauts. But he wrote one of the laws that explained how the balloon that came first was able to fly. His law varies temperature (T) and volume (V) while holding pressure constant. Charles' Law says:

The volume of a gas at a fixed pressure expands
in proportion to its absolute temperature.

The best example of this effect is a hot air balloon. Think of a paper grocery bag turned upside down over an oil lamp. Since the bag is open at the bottom, the pressure of the gas inside is the same as the pressure outside. However, the air inside expands when heated, and some leaks out, leaving fewer molecules of air inside the bag. The same is true for a hot air balloon, which also is open at the bottom. This makes a hot air balloon 'lighter than air', and Archimedes' Law – which will be covered later in this chapter - explains why the lighter bag of air can fly.

As mentioned earlier, Charles' Law is why upright freezers in a grocery store need doors, whereas tub freezers do not need lids. It is also why air density is reduced on hot days. Pilots call this **high density altitude**, and it is why airplanes need longer takeoff runs on hot days. They must go faster to push enough air molecules down. Mathematically, Charles' Law can be expressed either as $V_1/T_1 = V_2/T_2$ or $V/T = k$ where **k** is a constant number

If V/T is equal to a constant number, then when temperature (T) doubles, the volume (V) of that quantity of gas must also double. Does this mean that a temperature increase from 20°C (68°F) to 40°C (104°F) (which could occur between midnight and noon in the desert) would result in only half as much oxygen being available at noon? The answer is 'No'.

The catch is that ***absolute temperature*** does not double when temperature goes from 20°C to 40°C. The units are wrong. For the same reason, 104°F is not double 68°F.

To determine what 20°C vs. 40°C means in terms of *absolute temperature,* add 273 to degrees C and you will get degrees **Kelvin**. Accordingly, 313°K (i.e. 40°C +273) is only 6.8% higher than 293°K (20°C+273). ***Absolute zero*** is 0°K or -273°C.

Avogadro's Law

Amedeo Avogadro (1776-1856) put the jigsaw puzzle together by including the number of molecules **(N)** in gas equations. Avogadro's Law says:

Equal volumes of ideal gases at the same temperature
and pressure contain the same number of molecules.

i.e.   **N = k** when V, T and P are the same for the two gases

Bottom line, a one liter container full of hydrogen at room temperature and atmospheric pressure contains the same number of molecules as one liter of nitrogen gas at the same T and P. This is a little surprising, as H_2 has a molecular weight of 2, whereas N_2 has a molecular weight of 28. One might expect there to be 14 times as many of the much smaller hydrogen molecules.

Charles' Hydrogen Balloon relied on Avogadro's Law

"Equal violumes of ideal gases at the same temperature & pressure have the same number of molecules"

Molecular weight of H2 = 2

Fig. 2   Average Mol Wgt of air=28.6

However, because of their kinetic motion H_2 molecules essentially occupy the same volume of space as N_2 molecules. As a result, a bag of H_2 is much lighter than a bag full of air, which is heavier than pure N_2 because air is 77% N_2 and 21% O_2, and O_2 has a molecular weight of 32.

Avogadro's Law explains why Charles' balloon (Fig. 2) could be much smaller than the Montgolfier balloon and why humid air is lighter than dry air (H_2O molecules with a molecular weight of 18 replace heavier N_2 or O_2 molecules, respectively 28 and 32). Avogadro's Law also plays a role in the building codes that require a step up from the garage to the house.

Ideal Gas Law

The Ideal Gas Law combines Boyle's, Charles', and Avogadro's Laws into a single law that is expressed mathematically as:

$$PV = nRT \text{ where } R \text{ is a (constant) number.}$$

P, V, and **T** are pressure, volume and temperature respectively. The number of moles of gas is represented by **n**. One mole of gas contains 6.02×10^{23} molecules, so **n** is proportional to the number of molecules. Basically, **n** = the mass of the gas divided by its gram molecular weight.

Kinetic Theory of Gases

Gas molecules are in constant motion (Fig. 3), going faster when heated. This increases pressure on the walls of the container. Effectively, gas molecules "occupy" more space

Fig. 3	H₂	He	N₂
Molecular weights	2	4	28

than the molecule's physical size would predict, in the order of about 600 times at room temperature. Thus gases are compressible because the physical size of the molecule itself is minimal compared to the space it occupies.

Liquids Vs. Gases

Liquids and gases are both fluids (both "flow") but have a key difference: A liquid's molecules are in close contact with neighboring molecules, so liquids are essentially incompressible. This is why liquids work in hydraulic brake lines. As explained by the Kinetic Theory of Gases, gas molecules are much more mobile and have empty space between molecules. So gases are compressible, which is why air bubbles in a hydraulic brake line result in spongy brakes.

More importantly, a steam engine works because of the the massive increase in volume when heating liquid water causes it to transition to its gas phase, which we call steam. Further heating expands that gas. The first use in 1712 of steam engines was to pump water out of coal mines. About a century later, the automation of factories by steam engines, plus the introduction of steam trains and steam ships, brought about the Industrial Revolution.

Archimedes of Syracuse (c. 287 BC - c. 212 BC)

"Mr. Archimedes, you are going to get a reputation! Jumping out of the bathtub and running stark naked down the street yelling gibberish. What possessed you?" complained the housekeeper, as she handed the naked man a towel.

"I was yelling eureka. It's Greek, not gibberish", snapped the elderly scientist.

"Eureka" means "I have found it", and indeed he had. He had also established his reputation for the next few thousand years, although not for running down the street naked. While his bath time discovery had solved a problem posed by King Hiero ll of Syracuse, it also produced what is known in physics as Archimedes' Law:

> *Any object, wholly or partially immersed in a fluid, is buoyed up*
> *by a force equal to the weight of the fluid displaced by the object*

Since air and water are both fluids, this principle explains the buoyancy of boats, balloons and submarines. It was not his only contribution to aviation. Another, the Principle of the Lever, was key to how airplanes and dirigibles would be controlled two thousand years later.

King Hiero's Problem

King Hiero had provided a goldsmith with a quantity of pure gold to make a golden wreath. After the finished wreath was delivered, Hiero wondered if he was being cheated. Had the goldsmith substituted a lesser metal such as silver or lead for part of the gold? He challenged Archimedes to determine whether the wreath was pure gold or an alloy, but he could not destroy the wreath.

Chemical analysis was not an option, as the technology to do so was many centuries in the future. Density looked like it might provide an answer, as gold (density = 19.3 g/cc) weighs almost twice as much as either silver (density = 10.5 g/cc) or lead (density = 11.3). But how could he determine the volume of the wreath? Weighing it was less of an issue. He decided to relax in a tub of hot water and think about the problem.

Archimedes' inspiration came as he lowered himself into the tub. He noticed two things: (1) The surface of the water went up, and (2) he felt lighter. That is when he jumped out of the tub and ran down the street.

Archimedes determined the volume of the golden wreath by filling a tub with water and allowing the excess water to drain out

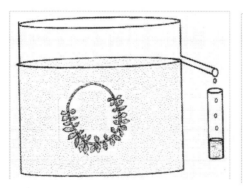

Fig. 4 Submerged object displaces its own volume in water

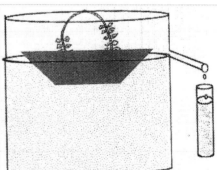

Fig. 5 Floating object displaces its own weight in water

through the overflow pipe. Then he carefully lowered the wreath into the tub, where it displaced its own volume of water (Fig. 4). The overflow was captured.

Next, he refilled the tub, placed a wooden bowl on the surface of the water, and allowed the displaced water to drain out. Then he placed the gold wreath in the floating bowl and captured the water displaced by the weight of the wreath (Fig. 5).

In Fig. 4 the fully immersed wreath displaced water equal to its own **Volume**. In Fig. 5, the floating wreath displaced its own **Weight** in water. If the water displaced in the 2nd experiment was 19.3 times the water displaced in the 1st experiment in either weight or volume, then the wreath was made of pure gold. If it surprises you that a heavy golden wreath would not cause the wooden bowl to sink, know that it would if the sides of the bowl were not high enough. Bottom line, the bowl and wreath combination will sink until it displaces its own weight in water. At that point, the buoyancy force pushing up will equal the downward pull of gravity.

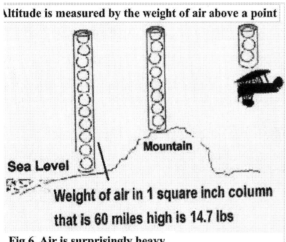

Fig 6 Air is surprisingly heavy

Similarly, the gravitational force pulling a hot air or hydrogen balloon down will be competing with the buoyancy force of the displaced air, which is pushing up. Importantly, it is not the amount of hydrogen you put in the bag that determines its buoyancy, it is the amount of air that is displaced by the bag. Accordingly, if the pressure of hydrogen inside the bag is the same as the ambient air pressure outside the bag, then the 'lift' will be at a maximum. Putting additional hydrogen into the bag under pressure only adds weight. In practice, balloons are not filled to their maximum volume in order to leave room for expansion as they go up in altitude where air pressure will be lower. As shown in Fig. 6 the pressure of the atmosphere at any altitude is the weight of the air above that altitude, so pressure drops as height above sea level increases.

In the case of a submarine, while its outside shape does not change, the contents of its ballast tanks do. When full of air, the submarine has positive buoyancy and floats. When the ballast tanks are flooded with water the submarine has negative buoyancy and sinks. To return to the surface, compressed air has to be blown into the ballast tanks to drive out the water. This makes the submarine lighter than the amount of water it displaces.

Falkirk Wheel

There are few better examples than the Falkirk Wheel for making two of Archimedes contributions to science more memorable. Together, Archimedes' Law and the Principle of the Lever allow enormous weights to be moved with a small expenditure of energy.

The Falkirk Wheel is an engineering marvel that replaced 11 locks in a canal system linking Glasgow on the west cost of Scotland with Edinburgh on the east coast. The challenge was to overcome an elevation difference of 79 feet between two canals. Boats enter either an upper or a lower tub. The doors are closed behind them and the two tubs plus all the water in them are exchanged by rotating the wheel.

Fig. 7 shows a boat entering the lower tub, while the upper tub is already loaded and closed. Each tub holds 300 tons of water when empty of boats. When the door closes behind a boat, the contents of the tub will still weigh 300 tons. This is because the floating boat displaces its own weight in water, per Archimedes' Law.

Add two more boats, or 20 people, and the weight in the tub will not change. More water will be displaced to reflect the increased weight that is floating per Archimedes Law. Meanwhile, the contents of the other tub will weigh 300 tons, regardless of whether on not there are any boats in it.

Fig. 8 shows the wheel in motion. The tubs are geared so that the top of each tub always faces upwards. Because the two tubs weigh the same, moving the combined 600 tons requires relatively little energy in accordance with the Lever Principle. The balanced system is rotated by a small motor generating only 22.5 KW, and completes one exchange in 15 minutes. For reference, the electric motor in a Toyota Prius produces 33 KW.

A key factor in keeping the tubs upright is the cogs on the smaller wheels in which each tub fits. These cogs ensure that the tubs are always upright, even though the frames within which these smaller wheels rotate must themselves rotate 180° when a boat goes from the upper canal to the lower canal.

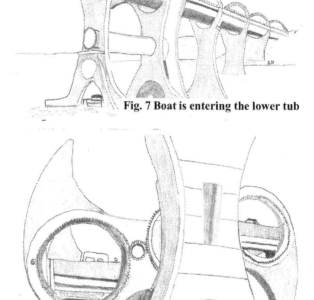

Fig. 7 Boat is entering the lower tub

Fig. 8 Wheel in Rotation

Bermuda Triangle & Archimedes Law

Another memorable application of Archimedes Law relates to the Bermuda Triangle. The Ocean Flatulence Theory for why ships disappear there relates to high concentrations of methane ("natural gas") under the ocean floor. If a large volume were to be released as a ship was passing overhead, then the ship would be floating on a bubbling mixture of methane and water. This would be much lighter than water alone and the ship should sink.

One TV special sought to disprove this theory by bubbling air under a sailboat. They had to add so much weight to the sailboat to make it sink, that they concluded that the ocean flatulence theory was a myth. However, I question their methodology.

The 'scientific method' involves advancing a hypothesis and then conducting experiments to prove or disprove it, and I saw two possible flaws in their experiment. One was that air has an average molecular weight of 28.6, whereas methane has a molecular weight just over half of that, specifically 16. Thus the air/water mix used in the TV special was heavier than a methane/water mix would be. Another possible flaw was the high volume and velocity of the air flow. When combined with the rounded shape of the sailboat's bottom, it may have been a source of lift. Examples of such lift include a beach ball kept in the air by a hair dryer, a freshly poured Diet Coke overflowing the glass in an airliner or why pouring Guinness is a 2-step process.

Chapter 4　Ben Franklin: Eye Witness to History

Ben Franklin had such an exciting time in Paris in 1783. He was there to negotiate the Treaty of Paris which ended the American War of Independence. An amazing man, he was the only one of America's "Founding Fathers" who signed all three of the key documents that created the United States. These were the Declaration of Independence, The Treaty of Paris, and the Constitution. Ben was also a well known scientist, or philosopher, as scientists were known in those days. Like everybody else in Paris, he knew about the two teams racing to be the first to fly a manned aircraft. That two week period began the Golden Age of Balloons.

A letter written by Ben Franklin after witnessing the ascent of Jacques Charles and Nicholas Robert communicates the excitement of the day:

TO SIR JOSEPH BANKS

Passy, Dec. 1, 1783. Dear Sir:-

In mine of yesterday I promised to give you an account of Messrs. Charles & Robert's experiment, which was to have been made this day, and at which I intended to be present. Being a little indisposed, and the air cool, and the ground damp, I declined going into the garden of the Tuileries, where the balloon was placed, not knowing how long I might be obliged to wait there before it was ready to depart, and chose to stay in my carriage near the statue of Louis XV., from whence I could well see it rise, and have an extensive view of the region of air through which, as the wind sat, it was likely to pass.

The morning was foggy, but about one o'clock the air became tolerably clear, to the great satisfaction of the spectators, who were infinite, notice having been given of the intended experiment several days before in the papers, so that all Paris was out, either about the Tuileries, on the quays and bridges, in the fields, the streets, at the windows, or on the tops of houses, besides the inhabitants of all the towns and villages of the environs. Never before was a philosophical experiment so magnificently attended.

Some guns were fired to give notice that the departure of the balloon was near, and a small one was discharged, which went to an amazing height, there being but little wind to make it deviate from its perpendicular course, and at length the sight of it was lost. Means were used, I am told, to prevent the great balloon's rising so high as might endanger its bursting. Several bags of sand were taken on board before the cord that held it down was cut, and the whole weight being then too much to be lifted, such a quantity was discharged as to permit its rising slowly. Thus it would sooner arrive at that region where it would be in equilibrio with the surrounding air, and by discharging more sand afterwards, it might go higher if desired.

Between one and two o'clock, all eyes were gratified with seeing it rise majestically from among the trees, and ascend gradually above the buildings, a most beautiful spectacle. When it was about two hundred feet

20

high, the brave adventurers held out and waved a little white pennant, on both sides their car, to salute the spectators, who returned loud claps of applause. The wind was very little, so that the object though moving to the northward, continued long in view; and it was a great while before the admiring people began to disperse.

The persons embarked were Mr. Charles, professor of experimental philosophy, and a zealous promoter of that science; and one of the Messieurs Robert, the very ingenious constructors of the machine. When it arrived at its height, which I suppose might be three or four hundred toises, [*A toise was a distance of about 2 meters*] it appeared to have only horizontal motion. I had a pocket-glass, with which I followed it, till I lost sight first of the men, then of the car, and when I last saw the balloon, it appeared no bigger than a walnut.

I write this at seven in the evening. What became of them is not yet known here. I hope they descended by daylight, so as to see and avoid falling among trees or on houses, and that the experiment was completed without any mischievous accident, which the novelty of it and the want of experience might well occasion. I am the more anxious for the event, because I am not well informed of the means provided for letting themselves down, and the loss of these very ingenious men would not only be a discouragement to the progress of the art, but be a sensible loss to science and society.

I shall inclose one of the tickets of admission, on which the globe was represented, as originally intended, but is altered by the pen to show its real state when it went off. When the tickets were engraved the car was to have been hung to the neck of the globe, as represented by a little drawing I have made in the corner.

I suppose it may have been an apprehension of danger in straining too much the balloon or tearing the silk, that induced the constructors to throw a net over it, fixed to a hoop which went round its middle, and to hang the car to that hoop.

Tuesday morning, December 2d.-I am relieved from my anxiety by hearing that the adventurers descended well near L'Isle Adam before sunset. This place is near seven leagues from Paris. Had the wind blown fresh they might have gone much farther.

If I receive any further particulars of importance, I shall communicate them hereafter.

With great esteem, I am, dear sir, your most obedient and most humble servant,

B. FRANKLIN

P.S. Tuesday evening.-Since writing the above I have received the printed paper and the manuscript containing some particulars of the experiment, which I enclose. I hear further that the travellers had *perfect command of their carriage, descending as they pleased by letting some of the inflammable air escape, and rising again by discharging some sand; that they descended over a field so low as to talk with the labourers in passing, and mounted again to pass a hill. The little balloon falling at Vincennes shows that mounting higher it met with a current of air in a contrary direction, an observation that may be of use to future aerial voyagers.*

My source of the above letter was Nathan G. Coodman, ed. *The Ingenious Dr. Franklin, Selected Scientific Letters of Benjamin Franklin* (Philadelphia: University of Pennsylvania Press, 1931), pp. 99-105.

Why were the balloons lighter than air?

Although Charles understood why his balloon flew, the Montgolfier brothers did not for their own balloon. They thought that it was the smoke that they created inside the balloon that provided lift. While much of the heated air was introduced into the balloon before they left the ground, there also was a frame inside the balloon on which they would burn straw and wool. Charles' hydrogen balloon was much more efficient, but hydrogen gas was both expensive to produce and a mystery as to what it really was.

While people knew that the gas in Charles' balloon was lighter than air, they had no idea as to its composition. Franklin called it "inflammable air" and knew that it was produced by reacting iron filings with sulfuric acid ("oil of vitriol" to Franklin). Today we would write the reaction as follows:

$$H_2SO_4 + Fe \Rightarrow FeSO_4 + H_2 \nearrow$$

Franklin called the gas in the Montgolfier balloon "air rarified by heat", suggesting that, unlike the Montgolfier brothers, he understood that it was expanded air rather than smoke that provided its lift. The science of chemistry was still in its infancy. Reading letters such as this one and the one that follows should be required reading in chemistry classes, as one can see what was in the minds of these philosopher scientists in the emerging days of the science. Not only is the historical perspective interesting, it often reveals the paradigms and fears that controlled their thinking.

In a December 1784 letter to John Ingenhousz (from *The Aeronautical Annual 1895* Edited by James Means) Ben Franklin talked about the different technologies in the two types of balloons and the potential impact of balloons on society in general, and war in particular.

Passy, January 16, 1784

DEAR FRIEND:

I have this day received your favor of the 2d instant. Every information in my power, respecting the balloons, I sent you just before Christmas contained in copies of my letters to Sir Joseph Banks. There is no secret in the affair, and I make no doubt that a person coming from you would easily obtain a sight of the different balloons of Montgolfier and Charles, with all the instructions wanted; and if you undertake to make one, I think it extremely proper and necessary to send an ingenious man here for that purpose; otherwise, for want of attention to some particular circumstance, or of not being acquainted with it, the experiment might miscarry, which, in an affair of so much public expectation, would have bad consequences, draw upon you a great deal of censure, and affect your reputation.

It is a serious thing to draw out from their affairs all the inhabitants of a great city and its environs, and a disappointment makes them angry. At Bordeaux lately a person pretended to send up a balloon, and had received money from many people, but not being able to make it rise, the populace were so exasperated that they pulled his house, and had like to have killed him.

It appears, as you observe, to be a discovery of great importance and what may possibly give a new turn to human affairs. Convincing sovereigns of the folly of wars, may, perhaps, be one effect of it, since it would be impractical for the most potent of them to guard his dominions. Five thousand balloons, capable of raising two men each, could not cost more than five ships of the line; and where is the prince who can afford so to cover his country with troops for its defence, as that ten thousand men descending from the clouds might not in many places do an infinite deal of mischief before a force could be brought together to repel them? It is a pity that any national jealousy should, as you imagine it may, have prevented the English from prosecuting the experiment, since they are such ingenious mechanicians, that in their hands it might have made a more rapid progress towards perfection, and all the utility it is capable of affording.

The balloon of Messrs. Charles and Robert was really filled with inflammable air. The quantity being so great it was expensive and tedious filling, requiring two of three days' and nights' constant labor.

It had a *soupape*, or valve, near the top which they could open by pulling a string and thereby let out some air when they had a mind to descend; and they discharged some of their ballast of sand when they would rise again. A great deal of air must have been let out when the landed so that the loose part might envelop one of them; yet the car being lightened by that one getting out of it, there was enough left to carry up the other rapidly. They had no fire with them. That is used only in M. Montgolfier's globe which is open at the bottom and straw constantly burnt to keep it up. This kind is sooner and cheaper filled, but must be of much greater dimensions to carry up the same weight, since air rarefied by heat is only twice as light as common air, and inflammable air ten times lighter. M. Morveau, a famous chemist at Dijon, has discovered an inflammable air that will cost only a twenty-fifth part of the price of what is made by oil of vitriol poured on iron filings. They say it is made from sea-coal. Its comparative weight is not mentioned.

I am, as ever, my dear friend,

Yours most affectionately,
B. Franklin

The "Inflammable air" produced from "Sea-coal" by M. Morveau was about 50% Hydrogen (H_2 with a molecular weight = 2), 35% Methane (CH_4 with molecular weight = 16) and 8% Carbon Monoxide (CO molecular weight = 28). "Sea-coal" was regular coal and got the name to differentiate it from charcoal because it was shipped by sea from Newcastle to London. So the average molecular weight of this gas mixture would be 8.8, just over four times heavier than pure hydrogen, but still only about 30% the weight of the same volume of air at the same temperature. Bottom line, it would have been lighter than the hot air used in the Montgolfier balloon.

For reference, "Air" is 78% Nitrogen (N_2 molecular weight = 28) and 21% Oxygen (O_2 molecular weight = 32), plus 1-4% water vapor (H_2O molecular weight = 18) and a trace of Carbon Dioxide (CO_2 molecular weight = 44), so the average molecular weight of air is approximately 28.6.

A final point on balloons: While helium has replaced hydrogen in gas balloons, and natural gas burners have replaced smoldering straw and wool in hot-air balloons, both types are still in use today. The 9-day annual Albuquerque International Balloon Fiesta attracts upwards of 700 balloons each October. These hot-air balloons are huge, gaudy aircraft, just like those of their pioneering predecessors in the 1780's.

Chapter 5   **Dirigibles – the First Airliners**

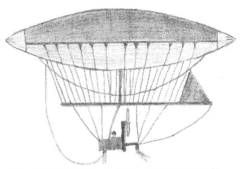

Giffard's Airship

Because they went only where wind blew them, early balloons were little more than hobbies or fairground attractions. Henri Giffard's dirigible (Fig. 1) changed this in 1852. He achieved the first powered, controlled, sustained flight, using a hydrogen balloon with a rudder and a propeller driven by a 3 hp steam engine.

The weight of the steam engine and the need to keep it away from the highly combustible hydrogen limited the

Fig. 1 Henri Giffard's Steam Airship (1852)

usefulness of Giffard's Airship. However, it pointed out what would be feasible when lighter sources of power became available in the form of diesel engines.

Zeppelins

Zeppelins were the first revenue generating airliners and the first long range bombers. The first one flew in 1900 (3 years <u>before</u> the Wright brothers!) and had two 14 h.p. Daimler engines. At 420 feet long, it was bigger than a B-747, which is 230 feet. A decade and a half later they were used as long range bombers in World War 1, as predicted by H. G. Wells in his 1907 book, *The War in the Air.* The last of the Zeppelin class was the airliner *Hindenburg* which went down in flames in 1937.

Fig. 2   First Zeppelin flight in 1900

In the early days of World War 1 Zeppelins flying at night faced little resistance when bombing Britain, as flying in the dark was a lot less challenging to an aircraft that could hang around until daylight to find a place to land. Unlike a toy rubber balloon that one blows up by mouth and has the air inside under pressure, the huge gas bags in a dirigible were at atmospheric pressure. Thus, prior to the introduction of incendiary tracer bullets, they had surprisingly little to fear from bullets going through the gas bags. The gas bags would not 'pop' like a child's balloon, and the resulting holes would vent a relatively small amount of H_2. In any event, it was still inside the envelope of the Zeppelin.

It is similar to the trick question: Would a B-747 with a million canaries on board weigh less if you could get all the canaries flying simultaneously? The answer is 'no', it would weigh the same. A canary weighing 2 ounces would need to push two ounces of air down to be able to fly, and that air would impinge on the floor of the plane. The force of two ounces of air would be the same as two ounces of sitting canary. Similarly, hydrogen that had leaked out of the gas bag into the outer shell of the Zeppelin would still have provided the same buoyancy force, since it continued to displace heavier air as predicted by Avogadro's Law.

The first Zeppelin destroyed by an airplane was actually brought down with bombs, not bullets. On June 6, 1915 Lt. R.A.J. Warnerford of the RAF intercepted Zeppelin LZ-37. After shooting at it for a while, with no effect observed, he climbed above the Zeppelin and dropped 6 bombs onto it from above and saw it burst into flames.

Although the Germans were the most successful builders of dirigibles/Zeppelins, Britain with its far flung empire was very interested right from the start. Several were also made in Italy and were used in polar expeditions.

The *Hindenburg* was the first airliner to provide regular non-stop air service between Europe and America, and took 2.5 days to cross the Atlantic at a time when even the fastest liners took five days and slower ships took ten. Its fastest crossing of the Atlantic was 42 hours and 3 minutes to go from Lakehurst NJ to Frankfurt, Germany. The luxury of this remarkable aircraft can be viewed at: http://www.theatlantic.com/infocus/2012/05/75-years-since-the-hindenburg-disaster/100292/

The four 16-cylinder Daimler engines drove 4-bladed, fixed pitch, 19.7' diameter propellers. The cruise setting of 1350 rpm generated 850 hp. *Hindenburg* was huge, 803.8 feet long and 135 feet in diameter, and containing over 7 million cubic feet of hydrogen. Surprisingly, they had to vent 1-1.5 million cubic feet of H2 during a trans-Atlantic voyage to compensate for the weight of fuel consumed during the flight. Otherwise it would have ascended too high.

Fig. 3 Demise of *Hindenburg* 1937

Originally designed to use helium (He) rather than H_2 for lift, 14 of the 16 gas cells were intended to hold inner gas cells of H_2 that could be vented rather than the much more expensive He. Thus the highly explosive H_2 would have been protected inside a larger cell filled with inert He. The USA was the sole source of helium at the time and, having declared it a critical war material, refused to sell any to Germany. So the *Hindenburg* had to rely solely on hydrogen as a lifting gas, setting the stage for the blazing demise of lighter-than-air airships (Fig. 3).

Hindenburg was the last of the Zeppelin class and is still the largest aircraft in history. The quavering voice of radio reporter Herbert Morrison ("Oh the humanity . . . ") as he narrated the horrible scene of the blazing dirigible smashing into the ground continues to have an effect. That effect is paradigm paralysis. Aircraft crashes are rarely filmed as they occur, and that widely distributed film with the accompanying dialogue ended the golden age of dirigibles on May 6, 1937. Even Hitler gave up on Zeppelins after that.

(To see the film of the crash and hear Herbert Morrison's horrified dialog, do a search on the Internet for "Hindenburg + Oh the humanity")

It would be a surprise to many that 61 of the 97 people on board *Hindenburg* survived. 13 of the fatalities were passengers. Significantly, they are the only passenger fatalities in the history of commercial dirigibles. Zeppelins were both the 1st revenue generating airliners, and the 1st trans-Atlantic airliners. *Hindenburg's* sister ship, *Graf Zeppelin* was 29 feet shorter and flew 1,053,391 miles, in its nine years. It carried over 34,000 passengers in 590 flights without a single passenger injury. On one around-the-world flight it flew from Friedrichshaven to Tokyo, then Tokyo to Los Angeles, Los Angeles to Lakehurst NJ, and finally Lakehurst to Friedrichshaven.

Surprising examples of Cause & Effect

Why were Germans banned from eating sausages during World War 1? Answer: Zeppelins.
The intestines of cattle were used to make the gas bags in Zeppelins, and a single airship needed 250,000 cows. As a result, the German government banned the use of cow intestines for making sausages and required butchers to hand over all that they harvested from slaughtered cattle.

During World War 2 the tires of many B-17s were pressurized with He to reduce weight and allow more fuel or bombs. This is in accordance with Avogadro's Law which states that *equal volumes of ideal gases at the same temperature and pressure will contain the same number of molecules.* The molecular weight of Helium is 4, so it is considerably lighter than N_2 molecules (molecular weight = 28).

The introduction of incendiary bullets caused the designers of Zeppelins to make the airframes lighter so they could fly higher, but that also made them weaker. After World War 1 the surviving Zeppelins were handed over to the Allies in war reparations, so much of the experience the Allies had in the post war years were with the lighter and weaker dirigibles, either built by the Zeppelin company or based on their plans. Such was the case with the US Navy's Shenandoah, which was built in the Philadelphia Navy Yard in 1919. At 680 feet long it was 10 meters longer than the German original, and was the first rigid dirigible to be filled with helium rather than hydrogen. It was the crown jewel of naval aviation until it was torn in half in a storm in 1925. A 600 foot long Zeppelin would be a lot more brittle than one with three 200 foot long tubes attached side by side. One wonders why they did not use the Aereon approach.

Aereon Hybrid Airship 1966

Solomon Andrews' gravitation powered balloon/dirigible in the 1860's was the obvious inspiration for Aereon lll (Fig. 4) a century later. Aereon lll was intended to be heavier than air and Aereon Corporation of Princeton NJ included other advances. The prototype was powered by a single propeller at the rear, and helium was to be heated in

Fig. 4 Aereon lll (1966)

the gas cells, thereby reducing the need to vent gas to descend, and making the Aereon lll a quasi-Roziere. Aereon lll was damaged in a taxiing accident and the company ran into financial problems.

The restructured Company pursued the Aereon 340 design (Fig. 5) and were awarded US Patent # 3,486,719 in 1969 for a hybrid of a rigid airship and an airplane. The dimensions can be judged by comparing the size of the nearby trucks and the aircraft in the drawings. Aereon 340 was to be 340' long with a 75' wing span. It was referred to as the 'Deltoid Pumpkin Seed' in a book of the same name by John McPhee. The patent described a gas filled delta shaped wing with internal cargo space sealed from the gas space. The objective was to get

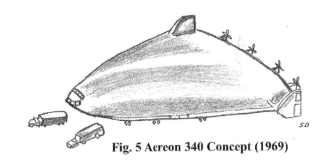

Fig. 5 Aereon 340 Concept (1969)

some lift from its aerodynamic shape and some from the gas, thereby overcoming disadvantages of both aerostatic airships and conventional cargo airplanes. Specifically, it would not need the high takeoff and landing speeds of the latter, while the ability to taxi on the ground would facilitate cargo loading and unloading.

The wheels would swivel so the aircraft could land with the aircraft pointed into the crosswind rather than straight down the runway similar to a B-52. The height of both the forward and aft undercarriage could be varied. For stability when parked the aircraft would be in a nose-down attitude similar to a Varieze in the grazing attitude. During the take-of roll the nose gear would be extended to provide a positive angle of attack, thereby facilitating flight into ground effect where ground friction would be reduced and further acceleration could take place. The design never made it beyond a small test prototype which managed to fly out of ground effect without gas bags installed.

Aeroscraft 2013

A half a century later Aeroscraft is a lighter-than-air airship concept that includes elements of all the earlier Aereon concepts and a submarine. It uses what its inventors call the COSH technology (Control of Static Heaviness). Buoyancy is controlled by a method similar to ballast tanks on a submarine, except that ambient air is used in place of water and compressed helium instead of compressed air. Buoyancy is achieved by releasing helium into bags that expand into empty space and displacing air. Compressing some of the helium back into compression tanks shrinks the bags and air then flows back into the now empty space, making the aircraft heavier again. This overcomes one of the major disadvantages of blimps. While they are great at going up, their huge size creates problems when landing. Without control over buoyancy, any large cargo that is off-loaded has to immediately be replaced with something of equal weight or the lift gas has to be vented.

The Aeroscraft's variable-buoyancy is expected to allow it to take off and land vertically without the need to exchange ballast. The goal is a cargo aircraft that does not require an airport or need a lot of ground crew and specialized facilities, so the U.S. Department of Defense, DARPA, and NASA have all partnered with the Worldwide Aeros Corporation in the development program. (For more detail on this emerging technology see http://aeroscraft.com)

The First Airplanes

Chapter 6

Fig. 1

The word "plane" means "flat surface" and describes the bottom of most wings, although there are variations. If its leading edge has a positive angle with respect to the relative wind (i.e. direction of flight through the air), then a wing's bottom will be the 'front' where air piles up, just as snow piles up in front of a snowplow (Fig. 1). That higher pressure provides most of the lift. The curved surface at the top of an airfoil increases the efficiency with which air is pulled down from above and 'behind' a wing, so it also provides lift. However, you could use a flat barn door as a wing if you could propel it fast enough. Some definitions are appropriate at this point:

- Angle of attack = Angle between wing chord line and oncoming air
- Wing chord line = Line from leading edge of wing to trailing edge
- Relative wind = Direction of flight = Direction of oncoming air

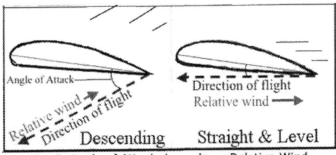
Fig. 2 Angle of Attack depends on Relative Wind

The first recorded use of a planar surface for flight was in China in the year 559 AD. Yuan Huangtou was one of 18 prisoners who were harnessed to thick bamboo mats and ordered to jump off the Tower of the Golden Phoenix. While the other 17 died when they hit the ground, Yuan managed to glide over the city wall and out into the countryside. Unfortunately, he was still harnessed to the mat and was recaptured. Since he was subsequently executed by some other means, I consider his flight more of a botched execution than a successful flight.

Who was the first to fly? Arab kids learn about Abbas Ibn Firnas who jumped off the minaret of Cordoba's Grand Mosque in Andalusia, Spain in 875 AD (Fig. 3). If you are wondering about there being a mosque in Spain in those days, you may also wonder why Osama Bin Laden's "Message to the Muslim People" included:

> "Let the whole world know that we shall
> never accept the tragedy of Andalusia."

Fig. 3 Ibn Firnas
(Baghdad Airport)

Bottom line, Muslim Moors from North Africa controlled most of Spain for about 800 years until they were expelled in 1492. That date should sound familiar to Americans, but back to Ibn Firnas: His landing was rough and severely injured his back. Since being able to walk away and do it again is a requirement for a 'successful' flight, I say Ibn Firnas does not qualify. Nevertheless, Baghdad's airport is named after him. If the statue outside the airport looks a little like Saddam Hussein, blame my drawing. (But it does look like him)

Hezarfen Ahmet Celebi's Hang-Glider 1630

Another Muslim aviator with an Airport named after him (in Istanbul) was Hezarfen Ahmet Celibi. It is claimed that he made the first inter-continental flight from Europe to Asia. He will be covered in more detail in the chapter "The Celebi Brothers Vs. Gravity".

Sir George Cayley's Governable Parachute 1853

In 1810 Englishman Sir George Cayley published a three-part treatise titled "On Aerial Navigation" which stated that Lift, Propulsion, and Control were necessary for a successful flight. He also described obtaining lift from an inclined plane, and the values of wing dihedral and camber. In the September 25, 1852 issue of *Mechanic's Magazine* he described his "governable parachute" and went on to have his coachman fly it the following year. Cayley was 85 at the time. Basically, the aircraft was a boat on wheels with sail oriented from side to side rather than high to low. On landing the coachman quit. He had been employed to drive a coach and not a flying machine. Thus, the first man to fly a plane and walk away was lost to history.

Fig. 4 Cayley's 'Governable Parachute'

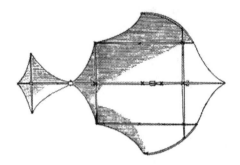

Otto Lilienthal's Hang-Glider 1895

Fig. 5 Lillienthal's Hang Glider

Prussia's Otto Lilienthal (1848-1896) may have been the first person to make repeated successful flights in a glider, although one source says that Hezarafen Ahmet Celebi did so. Lilienthal's was similar to modern hang-gliders and he was awarded U.S. patent #544,816 in 1895 for it. He extensively documented his flights and was an inspiration to the Wright brothers.

Because they always started at the top of a hill and ended up at the bottom, Cayley and Lilienthal get little credit inventing the "first heavier-than-air aircraft". It was fun but not very useful. The Wright Flyer overcame that deficiency in 1903 when it became the first heavier-than-air aircraft capable of taking off and landing at the same elevation. It met the requirements of a "good innovation", in that it was both useful and reproducible.

Wright Flyer: The Evolution of a Design

Orville and Wilbur Wright did not have a university degree between them, but they were not afraid to challenge the scientists and engineers of the day. However, they grew up a time when earning a high school diploma was more demanding than is true today. Avid readers of anything relating to aviation, they were also diligent experimenters.

The phrase "Invention is 10% inspiration and 90% perspiration" was certainly true in their case. They would develop a prototype, test it, refine it, and then test the new prototype in a sequential process that went on for years.

There were three distinct types of Wright Flyer with many intermediate steps between models. The rudderless 1900 Glider (Fig. 6) had the pilot lying prone, as did the 1903 Wright Flyer (Fig. 7) used in the historic first powered flight. The latter had bi-planar elevators and rudders extended forward and aft respectively. In the final 1911 Wright Flyer Model B (Fig. 8) the elevator was aft with the rudder, and the pilot and passenger were sitting upright.

Fig. 6 1900 Wright Glider   Fig. 7 1903 Wright Flyer   Fig. 8 1911 Wright Flyer Model B

The Wright brothers realized that they faced three hurdles: (1) Control, (2) Wing structure, and (3) Power. Their response was unique, and evidence that their design process was more evolutionary than revolutionary can be found in photos of the later models. In some cases it is clear that a particular flying machine started out as an earlier model, but had been modified to reflect the latest thinking, e.g. relocation of the elevators. Even their primary method of control, wing-warping, was "evolutionary" in that it mimicked the way birds control their own flight.

The rudderless 1900 glider had only wing-warping and elevator for control. They may not have even considered the need for a rudder, since birds manage to control flight by twisting their wings and tails and clearly do not have a vertical stabilizer or rudder. However, eventually the Wright brothers realized that control in three axes (longitudinal, vertical, and lateral) called for a control surface for each axis. The combination of wing warping, rudder, and elevator was their solution for the problem. Wing warping provided roll control around the longitudinal axis from nose to tail. The rudder at the rear provided yaw control around the vertical axis, and the elevator in front, which they called the horizontal rudder, provided pitch control around the lateral axis. U.S. Patent # 821,393 was granted in 1906 for a "Flying Machine".

This discussion brings to mind the question: "If a man can stand with just two legs, why does a stool need at least three?" The difference is that a man has a brain that rebalances him thousands of times a second, thus making a third leg unnecessary.

A good demonstration of this is to try and tie your left shoe while standing on your right foot and holding your left shoe off the floor. OK it's silly, but you will find yourself readjusting your balance repeatedly. An inanimate stool does not have a brain to control stability, and needs a third or fourth leg. Similarly, bird brains provide birds with adequate control without the need for a rudder. Move a feather here, a feather there, and the bird is balanced.

While the Wright patent described the technique of wing warping, the patent also claimed other methods for adjusting outer portions of the wings to achieve roll control. This is essentially what ailerons invented by Glen Curtis achieve.

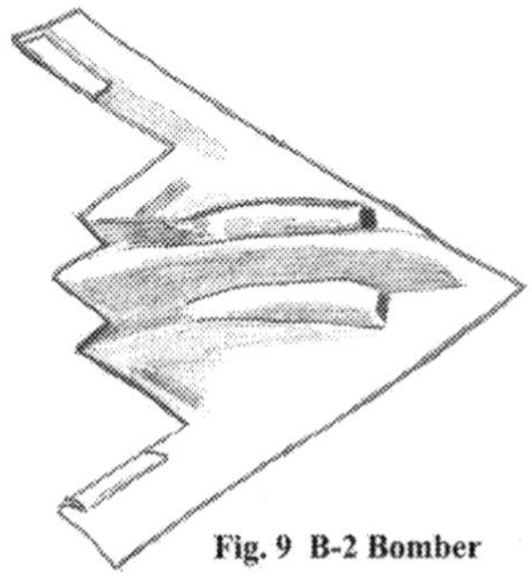

Northrup B-2 Bomber (1988)

It was not until the Northrup B-2 Bomber (Fig. 9) was built that a successful production airplane was designed without a rudder. Its computer

Fig. 9 B-2 Bomber

controls a combination ailerons, dive brakes, and flaps, in effect reshaping its wings just a bird does. Of course, such computer control comes at a price, $2 Billion per plane in the case of the B-2.

Elevator in Front – Cause & Effect

Back to the Wright brothers, they were the first to design 3-Axis controls, but why did they put the elevator in front in the early models? It would be nice to say that they got to make the rules because they were the first to play the game, or that they were balancing the rudder at the back. But it may have been more complicated than that. The 1900 model did not have a rudder, and its elevator was in front. They wanted to avoid the fate of Otto Lilienthal, who was one of their inspirations and was killed in 1896 when his hang glider stalled after hundreds of successful flights. It is possible that they saw the elevator in front as a shock absorber.

Bottom line, having the elevator in front limited their speed. The analogy "You cannot push a rope" applies. The rudder on a ship is always at the back for the same reason. If they made the elevator strong enough to stand up to higher speeds, they would have had to add too much weight given the materials available at the time. Theoretically, per the Principle of the Lever, a rudder could work equally well at the front. However, if the rudder breaks off, it will not work at all.

This weight Vs. strength issue was also why ailerons and rigid wings replaced wing-warping for control around the longitudinal axis, although Glen Curtis may have initially introduced ailerons as a way to bypass the Wright brothers patent. That was not a successful patent strategy. In any event, ailerons were a much better idea than wing warping, as it allowed the wings to be stronger. Control surfaces will be discussed in more detail in the next chapter which covers the physics of airplanes.

It is worth noting that the patent fight between the Wright brothers and Glen Curtis was to retard development of airplanes in the United States so much that the U.S. had no airplanes capable of entering World War 1. The aircraft that American flyers used in that war were all French. Eventually, the government had had enough and forced the aircraft builders to form a cross-licensing organization to last for the duration of the war. However, by the end of the war Wilbur had already died and Orville had sold his interest in the Wright Company.

Chapter 7

Sciences & Engineering of Airplanes

My favorite forum at the Oshkosh Fly In was where physicists and engineers would argue about which was a better explanation of an airplane's lift, Newton's Laws or Bernouilli Theory and the airfoil. From my perspective, both explain how planes can fly upright, but Newton's 3rd Law best explains how airplanes can fly upside-down.

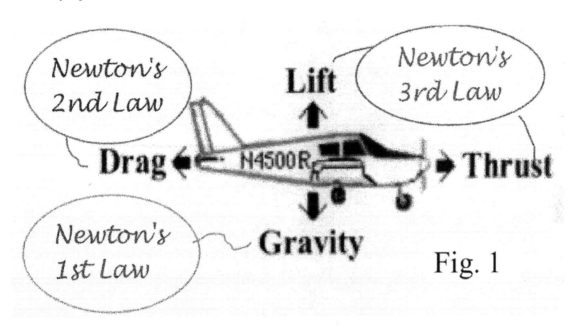

Fig. 1

The Four Forces

The Four Forces of Flight (Lift, Thrust, Drag, and Gravity) are all covered by Newton's Laws of Motion (Fig. 1). The force of gravity reflects Newton's 1st Law and the story of the

Newton's 1st Law of Motion	Newton's 2nd Law of Motion	Newton's 3rd Law of Motion
Objects at equilibrium will not accelerate	*Acceleration (a) is directly proportional to the net Force (F_{net}) and inversely proportional to its Mass (m)*	*Every action has an equal and opposite reaction*
Translation: *If you don't push, it won't move* and A moving object goes in a straight line until some external force causes it to turn, stop, or speed up	$a = F_{net} / m$ and $F_{net} = ma$ Translation: Larger Force→higher Acceleration Larger Mass→ lower Acceleration	Example: *Push air down and plane goes up* and *Push air back, plane goes forward*

apple falling on his head is how Newton explained it to people who were not 'philosophers', as scientists used to be known. It says that objects in a constant state of motion will remain in that state until influenced by an external force. In other words, they will not accelerate, slow down, or turn unless forced to do so. If they are motionless (i.e. acceleration = 0), then they will stay that way. One example is a golf ball on a tee waiting to be hit. Its "constant state of motion" (or equilibrium) is zero miles per hour, and it will sit there until an external force like a golf club or wind cause it to move. Similarly, a marble rolling across a table top will go farther than the same marble will go on a carpeted floor. The difference is the added resistance/drag/friction of the carpet, which is an "external force". A bullet fired in Space will continue will keep going until it hits something or enters a gravity field. One fired on Earth will slow down due to air resistance, and slope downwards because of gravity.

Newton's 2nd Law says the larger the mass, the greater the force required to accelerate or move it, and vice versa. Try hitting a golf ball and then a bowling ball with a golf club and you will understand what it means. Other examples include airplane drag and how a parachute works. Higher drag (i.e. air resistance) results in less acceleration in both cases.

$$a = F_{net} / m \quad \text{where } F_{net} = \text{Gravity - (Drag of parachute + man)}$$

The net force on a pilot coming down with a parachute is the force of gravity minus the force of drag caused by his body and the parachute. Thus, acceleration will be lower with a parachute than without. A toy rocket propelled by compressed air will not go as far as a rocket with both water and compressed air, because the mass of the water is greater than the mass of air.

Newton's 2nd Law also explains the difference between 'weight' and 'mass'. Mass *(m)* is the amount of stuff, and weight *(Fnet)* is the effect of gravity on that stuff, or *m* times *a*. Thus, a man with a mass of 90 kg on Earth will have a mass of 90 kg on the Moon, but will weigh only about a third as much because of the lower gravity on the Moon.

Blow up a balloon and release it and you have a perfect example of Newton's 3rd Law. Expelled air is the action and the balloon flitting around the room is the resulting reaction. Similarly, an airplane's propeller pushes air back, so the airplane goes forward. Lift is created by air being both pushed down by the bottom of the wing and pulled down by the top. As long as a wing has a positive angle of attack relative to the direction of motion through the air, its bottom will be the front where air piles up (just like snow in front of a snowplow).

If air was like jelly, then a wing moving through it would push all the air it hits downwards and leave a hole behind (Fig. 2). But air is not like jelly, and air would quickly fill the empty space from above. Thus, while the bottom (i.e. 'front') of the wing is pushing air down, the top (i.e. 'back') of the wing is pulling air down.

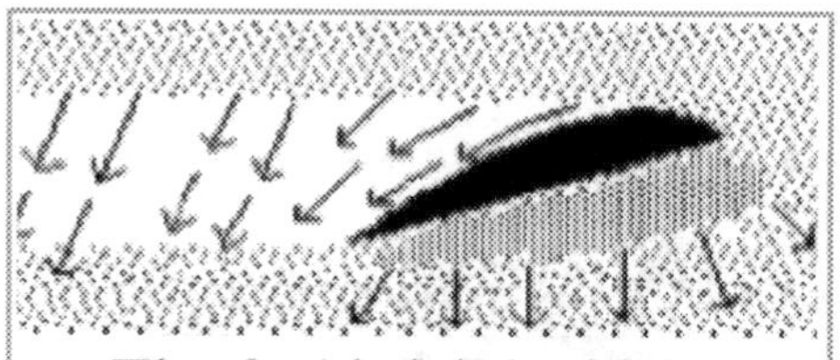

Bernouilli Principle – Low pressure on top of wing

Daniel Bernoulli (1700-1782) described the relationship between the velocity of a fluid and its pressure on whatever surface it is flowing over. The Bernouilli Principle states:
As the velocity of a fluid increases, the pressure exerted by that fluid decreases. Air going over the wing is much faster than air going under it. This is why many wings are given the **airfoil shape** shown in Fig 2, i.e. curved on top and flat on the bottom.

Bernoulli's Principle can easily be demonstrated by inserting part of a sheet of paper between the pages of a book and letting the rest of the sheet hang over, thereby simulating the top of an airfoil. Blow on the sheet of paper as depicted in Fig.3 and you will see the sheet rise up. Since the book prevents any air from getting under the sheet, the lift is provided entirely by air flowing over the top.

Fig. 2 Airfoil in flight

Bernoulli's Principle
Fig. 3

Coanda Effect: Air stays "attached' to top of wing

Henri Coanda (1886-1972) discovered that *when a jet stream of water or air comes into contact with a curved surface, it will attach itself to and follow the curve.* Blow out a candle held behind a soda can to demonstrate this effect (Fig. 4).

Dangle a spoon in a stream of water flowing from a faucet and it will be sucked in, and the stream will be deflected. This is a great demonstration of Newton's 3rd Law, Bernoulli Theory, and the Coanda Effect, where kids can both see and feel the physics involved.

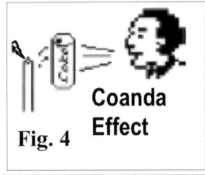

Fig. 4 Coanda Effect

'Stall' occurs when Bernouilli and Coanda fail

If the angle of attack is too great, air going over the wing cannot stay "attached" to the wing. It 'burbles' and no lift is generated by the top of the wing (Fig. 5). A stall can occur at any speed, although the most common times for accidental stalls are immediately after takeoff when the pilot raises the nose too much, or where a pilot has to make a tight turn on 'short final'. 'Low and slow' is a bad time to stall an airplane.

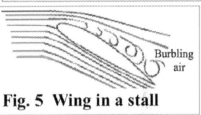

Burbling air

Fig. 5 Wing in a stall

To see other explanations and uses of the Coanda Effect, check out the following websites:
* Flying saucer: http://www.dragtimes.com/video-viewer.php?v=ggUIJDgkSSs&feature
* Surf boards: http://www.patentdata.com/report_full.php?aid=&pid=3318
* Avrocar: http://www.virtuallystrange.net/ufo/mufonontario/avro/avrocar.html
* Race cars: with inverted airfoils at the back to increase downward pressure on the rear wheels.

Airplane Control Surfaces

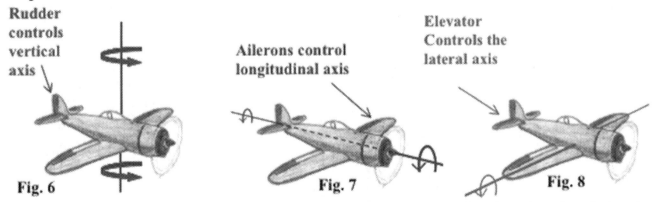

Rudder controls vertical axis

Ailerons control longitudinal axis

Elevator Controls the lateral axis

Fig. 6   Fig. 7   Fig. 8

Anybody who knows how to operate a see-saw will understand how the orientation of an airplane in flight is controlled, although it is complicated by the fact that an airplane has three fulcrums, while a see-saw only has one. Kids learn quickly learn how to make the see-saw go up or down. The control surfaces of an airplane (rudder, ailerons, elevator) work similarly.

An airplane's rudder (Fig. 6) works just like the rudder on a boat and airflow past it causes the stern to move left or right, while the bow – or nose on an airplane – goes in the opposite direction. In an airplane, this is called rotation around the vertical axis.

Airflow past the ailerons (Fig. 7) causes a plane to rotate around its longitudinal axis. Because ailerons move opposite to each other (i.e. when the left aileron is down the right one is up), they have opposing effects on a wing's ability to generate lift. With the left aileron down, curvature of the wing is increased and it has more lift. The opposite occurs on the right wing because its aileron is up. The net effect is the airplane rolls on its nose-to-tail longitudinal axis.

Finally, the elevator (Fig. 8), the Wright brothers "horizontal rudder", works similarly to the rudder but in a direction that is different by 90°. The lateral axis runs from side to side through the wings and the airplane is rotated around that axis by the elevator.

The flaps are the final control surfaces, and they are located on the wings' trailing edges like the ailerons, but are inboard of the ailerons. Unlike the ailerons, port and starboard flaps always go in the same direction. Lowered flaps cause two effects: (a) more lift due to the increased curvature of the wing and (b) more drag due to increased wind resistance under the wing. In effect, they work like built in parachutes, and allow to pilot to point the nose down (for landing) without causing speed to increase.

A common question is: Do all three axes intersect at one point, the c.g. (center of gravity)? The answer is 'Yes' and 'No'. Yes, they do intersect at one point, and no, it is not at the c.g., although it is close. The problem is that forces impinging on an aircraft sitting in a hangar are not the same as the forces impinging on it in flight. In the hangar, the only force acting on the plane is gravity. In flight the airplane has gravity <u>and</u> forces exerted by air moving over its surfaces, and these change with speed. This is why some specialized aircraft such as the F-14 Tomcat have variable wing configurations. Its wings are forward for the slower speeds of takeoff or landing, and swept back when it goes into combat. Bottom line, during flight an airplane rotates around its **aerodynamic center** (a.c.) rather than its **center of gravity** (c.g.).

In modern airplanes the c.g. is slightly forward of the aerodynamic center for a safety reason. If the engine quits, the pilot is trained to slow to the speed for maximum glide for that airplane and immediately start looking for a place to land. In that situation one does not want the aircraft to be tail heavy, as inattention could result in a flat spin that would be unrecoverable without the engine. (Remember how Goose was killed in the movie 'Top Gun'?). It is much safer for the plane to have a tendency to go 'nose down', as one can still use all of the control surfaces (ailerons, elevator, rudder, and flaps) as long as the airplane is going forward. Quite simply, the wings and control surfaces do not know if the engine is running.

Fig. 9   Rutan Varieze in Flight and "Grazing"

Having the c.g. slightly forward of the a.c. comes with a cost in increased drag. In order to neutralize a nose down tendency at cruise speed, the horizontal stabilizer (the flat surface forward of the elevator) and elevator are oriented so that they exert a downward pressure at cruise speed. The result is that an airplane weighing 2,000 lbs. needs to generate 2,200 lbs. of lift.

This Cause & Effect was brought home to me when I served as a time-keeper for an air race. As one plane was taxiing to the ramp I observed a woman in her mid-fifties climb out of the back seat into the front of a Cherokee. Their secret strategy was to takeoff with both women in front. One would then transfer to the back, thereby eliminating the need for the tail-plane to generate negative lift (and increase drag).

Burt Rutan's canard configurations (Fig. 9) bypass the problem entirely, as the elevator is at the front.

In both the flying and parked configurations, the nose wheel of a Varieze is retracted, for reduced drag and stability respectively. Without the weight of the pilot or passenger on board, the center of gravity of the airplane is far enough aft that a gust of wind could tip it over when the nose wheel is extended. Hence the unusual practice of retracting the nose wheel when parking the airplane, resulting in what is referred to as the "grazing" position.

Principle of the Lever

Archimedes made two major contributions to the sciences of flight. Archimedes' Law explained balloon lift, and the Principle of the Lever is used by aircraft control surfaces. He claimed that he could lift the Earth if he could find a lever that was long enough and a place on which to stand. He stated the Principle of the Lever as follows:

Fig. 10 Lever Principle

Magnitudes are in equilibrium at distances reciprocally proportional to their weights

That might have been clearer in Greek, but basically it means:

When a little kid and a fat kid are on a see-saw, the fat kid has to sit closer to the fulcrum

Continuing with the see-saw analogy in Fig. 10, the little kid and the fat kid are both being pulled down by gravity. Their respective weights are **FORCES**, and the distance between either of them and the fulcrum (i.e. point of rotation) is the **ARM.**

Multiply Force by Arm and the product is **MOMENT** (think 'momentum' and not 'minute'). Mathematically, the Lever Principle can be expressed as **Force x Arm = Moment**. Using a yard stick rotating around a nail at 18", and clamps at 15" and 21", the Arm on each side is 3" and the Force is 1 clamp. Thus, the moment on each side is 1 clamp x 3 inches = 3 clamp inches, and the yard stick is balanced, as shown in Fig. 11. However, if there were three clamps at 15" on the left side, then the single clamp on the right side would have to move to 27" for moments on both sides of the yard stick to remain balanced (Fig. 12).

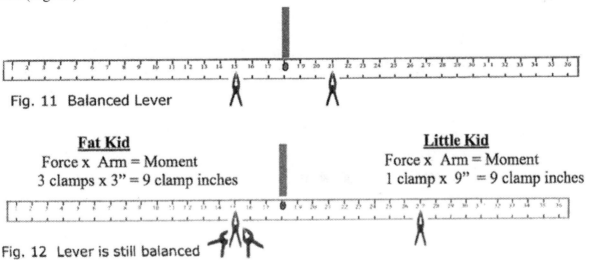

Fig. 11 Balanced Lever

Fat Kid
Force x Arm = Moment
3 clamps x 3" = 9 clamp inches

Little Kid
Force x Arm = Moment
1 clamp x 9" = 9 clamp inches

Fig. 12 Lever is still balanced

The problem with applying the Lever Principle to airplanes is that planes do not have a fixed fulcrum. Even the center of gravity is not fixed, as it changes when cargo is added or fuel consumed. The solution is to calculate moments from the nose of the airplane and include the weight of the empty plane. To illustrate how this works I weighed the clamps and the yard-stick. They weighed 15g and 35g respectively. Thus the yard-stick weighed 2.33 clamps, and its arm was 18 inches, the distance from one end to the middle of the yard-stick. Measuring from one end of the yard-stick you get the table shown below. Add up the weights and moments and then divide total moment by total weight to yield the center of gravity (c.g.). Not surprisingly, it turns out to be 18 inches. Weight & balance of airplanes is determined the same way.

Weight x **Arm**	=	**Moment**
3 clamps x 15 inches	=	45 clamp inches
1 clamp x 27 inches	=	27 clamp inches
2.33 clamps x 18 inches	=	41.94 clamp inches
6.33 clamps		113.94 clamp inches

$$\text{Center of Gravity} = \frac{\text{Total Moment}}{\text{Total Weight}} = \frac{113.94 \text{ C inches}}{6.33 \text{ C}} = 18 \text{ inches}$$

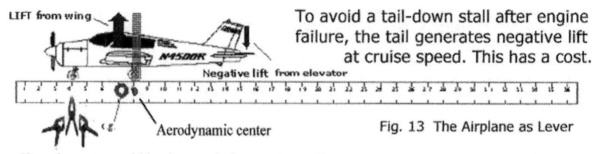

To avoid a tail-down stall after engine failure, the tail generates negative lift at cruise speed. This has a cost.

Fig. 13 The Airplane as Lever

How does one avoid having an airplane go 'out of balance' during a flight? Simple, have everything that will be consumed (fuel) or dropped out (bombs) close to the aerodynamic center.
As described earlier, an airplane in flight rotates around its 'aerodynamic center' rather than around its c.g. The relationship between the two (for most airplanes) is shown in Fig. 13.
    While a Varieze (Fig. 9) gets most of its lift from its wings, it gets some lift from the canard elevator in front. The canard is oriented so that it will stall at a higher airspeed than that at which the wings stall. Thus the nose will drop if the airplane slows down after engine failure.
However the wings are still generating lift and the nose-down attitude will cause speed to increase, effectively making the Varieze stall-proof. Effective use of Cause & Effect analysis made the Varieze safer, more fuel efficient, faster.

Air Pressure

Although gravity is pulling straight down, air pressure on all sides will hold a loose lid on a nearly full, but upside-down, glass of water

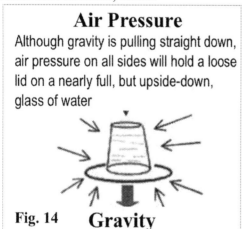

Fig. 14   Gravity

Air has weight and pressure

    You can no more feel the weight and pressure of the air around you than a goldfish knows that it is surrounded by water. The reason is that the air has always been there for you. When a goldfish jumps out of the bowl, it quickly finds what it had taken for granted. If you were suddenly ejected into the vacuum of Space, your eyes would bug out, your blood would begin to boil, and your brains might start to ooze out through your nose. Not a nice image, but certainly memorable, and that is key to learning anything. You need to remember it.

Standard atmospheric pressure at sea level is 14.7 psi. (psi = pounds per square inch). That pressure is all around you and not just on your head, a fact that can be demonstrated by putting the lid from a margarine tub loosely on top of a glass of water. Turn the glass of water upside-down while holding the lid, and then let go of the lid (Fig. 14). It appear as if the water in the glass is defying gravity. What is actually happening is that air pressure, helped by water's surface tension, is holding the lid onto the glass.

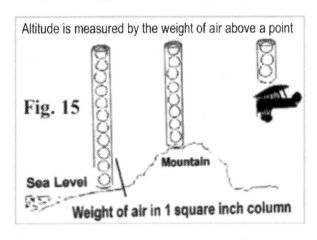

If you are 60 miles above the Earth's surface, you would be in Space and Earth's atmosphere would all be below you. Furthermore, if you erected a vertical 60 mile long tube with a 1 square inch cross section next to the ocean, the weight of air inside the tube would be 14.7 lbs.

Air pressure is how altitude is measured, but one has to correct for the barometric pressure of the day. Using the above 1 sq. in. column analogy, the column above a mountain would be shorter than one starting at sea level, and that above an airplane flying over the mountain shorter still (Fig. 15). Thus, the weight of the air in these columns and the pressure they exert would be correspondingly lower.

On a hot day, air in these imaginary columns would expand in accordance with Charles' Law and some would spill out of the column. Even though the pressure would be the same, the weight of the air remaining inside would be lower, as would its density. This lower absolute density is referred to as higher **density altitude** by pilots and has to be taken into account when determining takeoff and landing distances. Density altitude will be covered in more detail in the chapter 'Airplane Math'.

Wing configurations

Some pilots like high-wing airplanes, with the wings above the cockpit, others prefer low-wings. The advantages and disadvantages to each configuration can be understood from the collage in Fig. 16

A Cherokee has a lower center of gravity (c.g.) and a wider wheel base. Kids get it right away. When asked which can go faster around a corner, a Big Wheels or a traditional tricycle, kids will immediately respond with the former. Similarly, the wider wheel base and lower c.g. of a Cherokee is more reassuring when landing or taxiing with a crosswind.

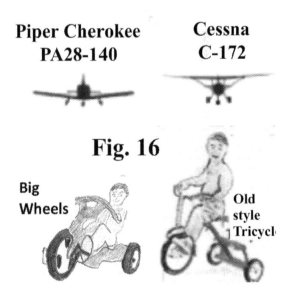

However, a Cherokee's low wing can give one pause when landing on a narrow runway with snow banks on each side. I once landed my Cherokee behind a C-172 on a narrow grass runway with corn growing on each side. I had to be precise in touching down in the center of the runway. My friend in the C-172 could be off on either side as his wings were above the corn.

The following table illustrates these points further and adds motivational dimensions to the analysis. (I am trying to make this scientific)

Advantages of Low Wing Configuration	Dim	Disadvantages of Low Wing Configuration	Dim
1. Easier crosswind landings due to wide wheel base & low center of gravity.	FAS	1. Wing gets in the way when you want to look at Grandma's house	FAO
2. Can see runway in turns to Base & Final	FAS	2. Not great for photography; Can't open window and wing gets in way	FAO
3. Easier preflight inspection gas, flaps, etc.	PAS	3. Door on passenger side only (usually).	PAO
4. Takes a klutz to bump head on wing.	PAS	4. Low wing precludes parking anything under wing.	PAO
5. Pronounced ground effect helps takeoff	FAS		

Advantages of High Wing Configuration	Dim	Disadvantages of High Wing Configuration	Dim
1. Easy to point out Grandma's house while in flight.	FAO	1. Crosswind landings can be dodgy due to narrow wheel base and high center of gravity.	FAS
2. Doors on both sides of aircraft.	PAO	2. Lose sight of runway in turns to Base & Final	FAS
3. Gravity fed fuel flow from wing tanks.	FAS	3. Everybody bumps head on wing sooner or later.	PRS
4. Park car under wing in hangar.	PRO	4. Certain to flip over in wheels-down water landing.	FRS
5. Can sleep under tarp slung over wing during the Oshkosh Fly-In	PRO	5. Must be 8' tall or a gymnast to check fuel during preflight inspection if ladder not available.	PAO

The 3-letter symbols by each advantage or disadvantage represent the motivational dimensions of each rating. They are Flying vs. Parked, Always vs. Rarely, and Safety vs. Other. Thus, FAS means (a) Flying, (b) Always happens, and (c) Safety related. PRO means the plane is Parked, the advantage or disadvantage Rarely happens, and something Other than safety is involved.

Obviously, FAS should motivate more than PRO. The other combinations lie somewhere in between these two. Highest priority for FAS is my paradigm and yours may be different. If you are one of the toy-mad types *("He who dies with the most toys wins")* and need to park a sports car under the wing to feel fulfilled, then perhaps PAO is your top priority and you need a high wing airplane.

Although a Cherokee wins for stability on the ground, theoretically a C-172 might have an advantage in the air. Most of the weight of the Cessna hangs below the source of lift, the wings, whereas most of the weight of a Cherokee is above the wings. This difference is reflected in the differences in **wing dihedral** between the two aircraft.

Wing Dihedral

Wing dihedral is the upward sweep of the wings from the fuselage to the wing tip, and represents an advanced application of the Lever Principle. Low wing airplanes (i.e. wings below the fuselage) like Piper Cherokees benefit significantly from dihedral, whereas high wing airplanes such as those made by Cessna have – and need – less dihedral. Fig. 17 shows how to explain dihedral to kids. The arrows show the direction of lift, which is 90° from the wings surface, and the plumb bobs show the direction of gravity,

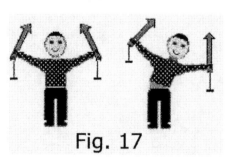

Fig. 17

which is always straight down. Although some of the wing's lift is lost because it does not directly oppose gravity, the effect is the same on both sides as shown in the figure on the left. This represents an airplane flying with its wings level. However, look at the difference between the two wings when one is dipped, as might happen in turbulent air. All of the lift of the lowered wing is directly opposed to gravity, whereas only a fraction of the lift of the raised wing opposes gravity. The net result is a tendency to return to level flight and more stability.

In a perfect example of Cause & Effect analysis, **Charles Lindberg** exploited this idea by having the ***Spirit of St. Louis*** built with zero dihedral. Flying solo from New York to Paris, he needed to stay awake at all times. Lack of dihedral kept him busy throughout the flight. By understanding why dihedral had been introduced into airplane design in the first place (i.e. greater stability), Lindberg was in a position to sacrifice that benefit for what he knew would be either a record breaking flight or his epitaph.

Chapter 8 STEM Q&A - Chemistry & Physics

This chapter uses a Q&A format to review and expand on chemistry and physics already covered in the book. Cover the right column with a sheet of paper and try to answer the questions. Then slide the paper down and check your answers. If you don't cheat, it is a great way to learn. If it is boring, then you have already learned it. Congratulations!

Questions	Answers
LIFT of a hydrogen or helium balloon is explained by (a) Boyle's Law? (b) Charles' Law? (c) Avogadro's Law?	**(c) Avogadro's Law"** *Equal volumes of Ideal Gases, at the same temperature and pressure, contain the same number of molecules* Molecular Weights: $H_2 = 2$ $N_2 = 28$ $O_2 = 32$ Thus, a hydrogen balloon is lighter than one filled with air
What law(s) of science explain why planes need longer takeoff runs on hot days? (a) Newton's 1st & 3rd Laws (b) Newton's 3rd & Charles' Law (c) Newton's 3rd Law & Bernouilli's Principle (d) Charles' and Avogadro's Laws	**(b) Newton's 3rd Law and Charles' Law** Air density will be lower the hotter air gets per Charles' Law (i.e. the air expands and there are fewer molecules). Thus the plane has to go faster to push enough air down and generate lift per Newton's 3rd Law.
At constant temperature & pressure, what effect will higher humidity have on the takeoff run? (a) A longer takeoff run will be required (b) A shorter takeoff run will be required (c) Higher humidity will have minimal effect	**(a) A longer takeoff run will be required** Avogadro's Law says that the number of molecules will be the same, so some N_2 or O_2 molecules will be displaced by H_2O molecules, which are lighter than the other two. Molecular weights: $O_2 = 32$ $N_2 = 28$ $H_2O = 18$
When gas temperature goes from 20° to 40°, does its volume double per *Charles' Law? (a) Yes (b) No ** At constant pressure a gas expands in direct proportion to its absolute temperature*	**b) No** The catch is "absolute" temperature, °K. To determine this, add 273 to degrees °C. Thus $40°C$ ($313°K$) is only 6.8% higher than $20°C$ ($293°K$), and not double.
Which is easier to make, Hydrogen or Helium? (a) Hydrogen gas (H2) (b) Helium gas (He)	**(a) Hydrogen gas (H_2)** H_2 is easy to produce, e.g. with a battery and water, or iron filings and any acid. Helium is a Noble Gas. It combines with no other atoms, and takes a nuclear reaction to produce. Helium released from balloons is so light it will go to high altitudes and effectively be lost in Space

Questions	Answers
Airplanes stall when which of the following happens? (a) Air ceases to flow smoothly over top of wings (b) Air ceases to flow smoothly under the wings (c) The engine quits (d) Airspeed drops below prescribed level	**(a) Air ceases to flow smoothly over top of the wings** When air 'burbles' over the top of a wing, no lift is generated by the top of the wing, although its bottom continues to produce lift by pushing air down.
Airplane control surfaces work by means of what laws of science? (a) Newton's 3rd Law (b) Principle of the Lever (c) Bernouilli's Principle (d) Newton's 3rd Law & Lever Principle	**(d) Newton's 3rd Law & Lever Principle** *Every action has an equal and opposite reaction* and Moment = Force x Arm Control surfaces (aileron, rudder, and elevator) capture the force of air flowing past them in flight
Which Control Surface rotates an airplane around its lateral axis? (a) Elevator (b) Ailerons (c) Rudder	**(a) Elevator** 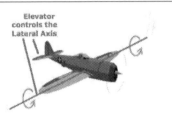
Which Control Surface rotates an airplane around its vertical axis? (a) Elevator (b) Ailerons (c) Rudder	**(c) Rudder** 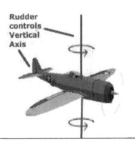
What law of science explains the LIFT of a hot air balloon? (a) Newton's 3rd Law (b) Avogadro's Law (c) Charles' Law (d) Clark's 3rd Law	(c) Charles' Law *At constant pressure, the volume of an ideal gas expands in direct proportion to its absolute temperature* Note: Hot air balloons are open at the bottom, so the pressure inside is the same as outside
Which law of physics explains how a parachute works? (a) Newton's 1st Law of Motion (b) Newton's 2nd Law of Motion (c) Newton's 3rd Law of Motion (d) The Principle of the Lever	**(b) Newton's 2nd Law of Motion** a = Fnet / m and Fnet = Gravity - Drag Since the parachute increases Drag, and the mass (m) is constant, then acceleration (a) will be lower.

Questions	Answers
Planes need longer take-off runs on hot days. Which of following will be true about the landing roll on a hot day? (a) The landing roll will be longer (b) The landing roll will be shorter (c) It will be the same as on a colder day	**(a) The landing roll will be longer** Per Charles' Law, air density is lower on a hot day, so planes have to go faster to push the same number of air molecules down. A faster airplane will require more runway to land. Short runways are challenging on hot days. The same is true for runways at high elevations where air is less dense.
Henri Gide's 1852 airship used a steam engine. What law(s) of science are involved in steam engines? (a) Charles' and Avogadro's Laws (b) Boyle's and Avogadro's Laws (c) Charles' and Boyle's Laws (d) Archimedes Principle	**(c) Charles' and Boyle's Laws** Charles' Law: *At constant P, a gas expands in direct proportion to its absolute temperature* Boyle's Law: *At constant T, V is inversly proportional to P*
A Hydrogen molecule (H_2) contains two atoms, whereas Helium (He) has only one atom. Which is lighter, H_2 or He? (a) Hydrogen (H_2) (b) Helium (He)	**(a) Hydrogen (H_2)** Hydrogen has 1 proton and 1 electron. Thus H_2 has a molecular weight of 2. Helium has two protons, 2 electrons, and two neutrons, so its molecular weight is 4. Electrons are light enough relative to protons and neutrons to ignore.
Why do Piggly Wiggly's tub freezers not need a lid? (a) Boyle's Law (b) Charles' Law (c) Avogadro's Law (d) Kinetic Theory of Gases	**(b) Charles' Law** Per Charles' Law, gases expand with rising temperature, so density goes down and the volume of gas weighs less. The opposite is also true. Cold air descends and stays in the tub freezer. Vertical freezers need a door.
Which Control Surface rotates an airplane around its longitidunal axis? (a) Elevator (b) Ailerons (c) Rudder	**(b) Ailerons**  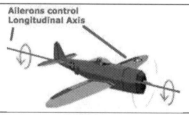
The best explanation of how an airplane can fly upside-down is provided by which of the following? (a) Bernouilli Principle (b) Newton's 3rd Law of Motion (c) Coanda Effect (d) Newton's 2nd Law of Motion	**(b) Newton's 3rd Law of Motion** *Every action has an equal and opposite reaction* Air is pushed down by the bottom (i.e. front) and pulled down by the top (i.e. back) of the wing 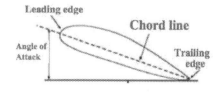
"Angle of Attack" describes which of the following? (a) Angle between relative wind and wing chord line (b) Angle between horizon and wing chord line (c) Angle of wing's bottom and the horizon (d) Angle - up or down - of the elevator	**(a) Angle between relative wind and wing chord line**

44

Questions	Answers
What forces influence an airplane in flight? (a) Lift, Drag, and Thrust (b) Air flowing over and under airplane's surfaces (c) Air pressure & air flowing over & under plane (d) Gravity, Thrust, Lift, and Drag	**(d) Gravity, Thrust, Lift, and Drag** 
Which of the following statements is true? (a) On Earth, weight and mass are the same (b) An object's mass on Earth is the same on the Moon (c) The mass of an object is different under water (d) Airplanes get more lift on humid days than dry days	**(b) An object's mass on Earth is the same on the Moon** Basically, 'mass' is the amount of stuff, and weight is the effect of gravity on that stuff. Thus, while mass is a constant, weight will be different on Earth Vs. the Moon.

Chapter 9   The Celebi Brothers Vs. Gravity

"So, your donkey will not run away unless it is the will of Allah that he do so. Of course, it would do no harm to close the gate"
. . . . Mehmet added with a smile.

Fig. 1 Lagari Hasan Celebi

The Harvard education of a Muslim friend revealed itself as he explained to me that nothing happens unless God was willing to have it happen. This is why Muslims often add *insha'Allah,* meaning *"If it is the will of God,"* when describing what they intend to do. He speculated that this attitude might be why Muslim nations are less developed technologically. While they have their share of scientists, too much of their populations wait for God to make something happen rather than taking the responsibility to do it themselves. A case in point is the Celebi brothers, Hezarfen Ahmed and Lagari Hasan Celebi, who reportedly made the first intercontinental flight and first manned rocket flight 150 years before the Montgolfier brothers and 270 years before the Wright brothers. Why was there no follow-up to their breakthrough flights?

I had never heard about Hezarfen until I found him while doing research on Lagari Hasan Celebi, the first survivor of a manned rocket flight. That occurred in 1633 AD in Constantinople. Watching were Sultan Murad IV, a few wives and his court. Fig. 1 depicts Celebi's rocket. A model of it can be seen in the Topkapi Museum in Istanbul, Turkey.

The *Crash & Burn* episode of the TV show *Myth Busters* attempted to replicate Celebi's rocket flight. The only contemporary report said that he attained an altitude of 1,000 feet before descending into the waters of the Bosporus "on the wings of eagles". Were these "wings" a parachute or actual wings as in a wing-suit or a hang-glider? That was what I was trying to find out. Unlike modern day wing-suit flyers, who have to deploy a parachute to land safely, Lagari Hasan Celebi planned to land on water. So he had more options to slow his descent after ejecting from the rocket.

Fig. 2 Hezarfen Ahmet Celebi

A snippet (http://www.youtube.com/watch?v=Z48P9VQJIhU) from the movie *Istanbul Beneath My Wings* shows a cross between a parachute and a wing-suit, but I would not rule out folding hang-glider wings. A depiction of his brother Hezarfen's hang-glider is shown in Fig. 2.

Hezarfen Ahmed Celebi used a hang-glider three years earlier to make the first inter-continental flight. That was not a great as it sounds, as he merely flew from one side of a city to the other. However, the city was Constantinople and the take0ff point was the Galata Tower on the European side, and the landing was in the Dogancilar Square in Uskudar on the Asian side of the Bosporus. A quick geography and history lesson is in order before getting into the details of Hezarfen's flight.

On the west side of the Sea of Marmara lie the Dardanelles, a narrow strait which leads into the eastern Mediterranean. The Bosporus is the narrow strait at the east end of the Sea of Marmara which connects it to the Black Sea and separates Europe from Asia. That is where the ancient city of Constantinople lies. The city was named after the Roman Emperor Constantine, who made Christianity the official religion of the Roman Empire in 313 AD and who moved the capital of the

Empire to Constantinople in 330 AD. It remained a Christian city until it was captured by the Ottoman Empire in 1453. The siege included the first use of canon guns, which in turn triggered one of the great transitions in history (This will be discussed in the chapter titled "Mega-historical Cause & Effect"). Constantinople was renamed Istanbul in 1923 after the demise of the Ottoman Empire following World War 1.

Hezarfen Celebi, for whom one of the three airports in Istanbul is named, was inspired by the 875 AD flight of another Muslim aviator, Abbas Ibn Firnas, after whom the airport in Baghdad is named. Firnas flew a hang-glider off the tower of the Grand Mosque in what is now Cordoba, Spain. However, his flight cannot be considered a success, as he had to be carried away afterwards and never walked without a crutch again. In contrast, Hezarfen must have made multiple flights before venturing across the Bosporus.

While I have described Hezarfen's aircraft as a hang-glider, it might be more accurate to call it a soaring glider. My reason for saying this is the glide ratio required for his trip across the Bosporus, which I calculate to be 39:1. That is high for a hang-glider, so some soaring was required. The distance from the Galata Tower to Dogancilar Square is just over two miles, 11,673 feet to be precise (Fig. 3). The top of the Galata Tower is 334 feet msl, and the elevation of Dogancilar Square is 39 feet msl, yielding an elevation change of 299 feet. (Thank you Google Earth and various Internet websites) Presumably, he picked up an up-draft after departing the tower and set out across the strait when he had enough altitude.

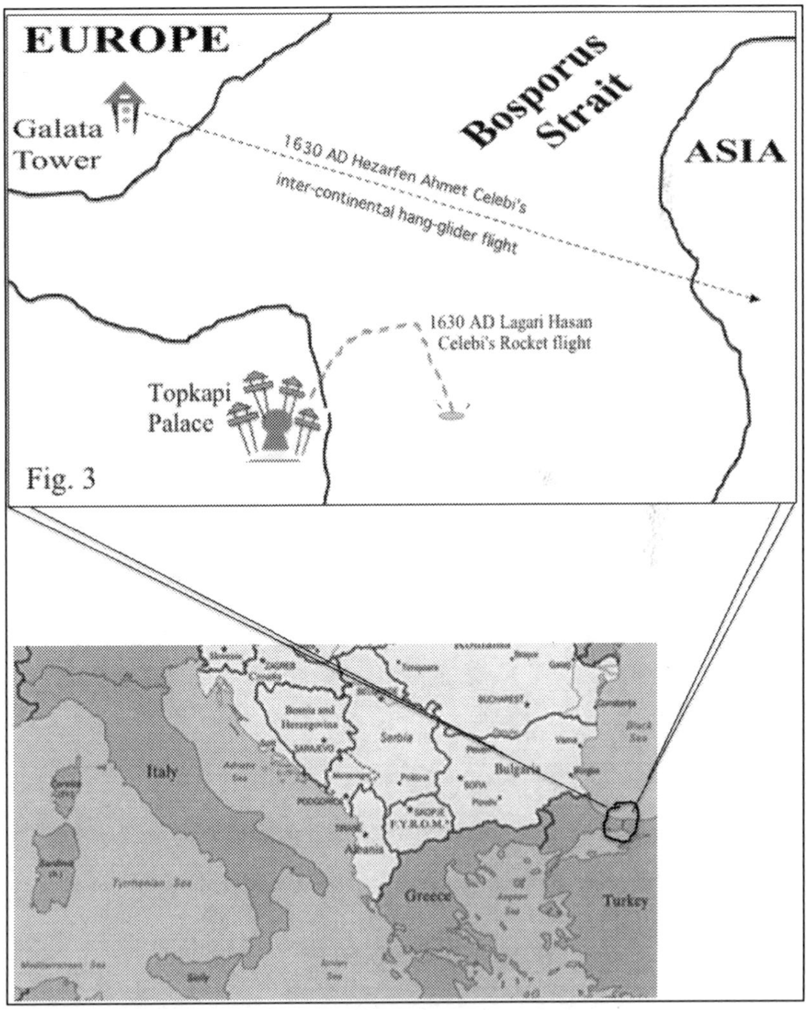

Fig. 3

Soaring gliders can stay aloft as long as the pilot can find an updraft. The record long distance flight in a glider was set in 2009 in New Zealand by Terry Delore and John Kokshoorn. It amounted to 1,554 miles. However, they flew along a mountain range which had an updraft on the upwind side. Hezarfen Ahmet Celebi was flying over a body of water, so updrafts would be less powerful, if they existed at all.

Back to the question I asked earlier, why were the breakthrough flights of the Celebi brothers lost to history? What Cause & Effect was in play? While both men were rewarded for their accomplishments by the Sultan, both were sent away from Constantinople soon after. One reported reason was "it was dangerous to have such men around". Did they rock the boat too much with their breakthrough technologies? As Galileo had found out a century earlier in Rome, challenging conventional wisdom could come with a price. Fortunately, western countries got over this hang-up and created the Industrial Age and all that followed.

One of George Bernard Shaw's *Maxims for Revolutionists* comes to mind:

> *The reasonable man adapts himself to the world; The unreasonable one*
> *persists in trying to adapt the world to himself. Therefore all progress*
> *depends on the unreasonable man.*

In contrast with the complacency often implied in the phrase *insha'Allah,* America is fortunate that an attitude of self-reliance prevails. It is one result of the pioneering spirit that drove many of our ancestors to leave the class systems of Europe and Asia behind when they ventured to the New World and it remains to be seen if the growing nanny state and reliance on big government will sap this spirit of self-reliance. Cause & Effect works in both directions.

Chapter 10 Rockets – Some Manned, Some Mean

Balloons get lift by being lighter than air. Airplanes get lift by both pushing and pulling air down by means of inclined planes (i.e. wings) that are being propelled through the air. Rockets lack such finesse. They are brute force, action/reaction per Newton's 3rd Law.

When Venetian explorer Marco Polo went to China in 1265 he saw gun powder and rockets. Several hundred years passed before there was a record of anybody using rockets to fly. Why so long? Was part of STEM in short supply?

In 1500 AD an astrologer by the name of Wan Hu wanted to go to the moon. He must have been a nut-job, because he attached 47 of the biggest rockets that he could find to a sturdy chair. He then ordered 47 servants to light all of the rockets simultaneously while he sat on the chair!

The servants did so and ran for their lives. A great bang ensued and there was much smoke. When that cleared, Wan Hu and the chair were nowhere to be seen. Did he make it to the moon? Who knows? But I doubt it.

First Rocket-man (1633 AD)

The previous chapter described the story of the Turk named Lagari Hasan Celebi who launched himself with rockets in 1633 and descended using "eagle-like wings" into the Bosporus. The lack of any follow-up to this feat always amazes me.

Fig. 1 Wan Hu's Rocket Chair

Robert Goddard

Although he got little attention in his own country in the late 1920's, Robert H. Goddard of the USA gets the credit for the first rocket with liquid rather than solid fuel. He used gasoline and liquid oxygen. Unlike solid fuel rockets, liquid fuel rockets can be throttled up or down, or shut off and restarted. There were a lot of similarities between Goddard's rockets and the Nazi V-2.

First into Space

The *V-2* rocket entered the war too late to make a difference for Germany. Climbing to a height of 61-63 miles, it was the first man-made object into Space, which begins 60 miles up.

Rocket Planes

The Me-163 *Komet* introduced by Germany in WW2 had limited success because of the short duration of its powered flight (6 minutes) and the hazardous nature of its fuel, T-Stoff (hydrogen peroxide) and C-Stoff (hydrazine hydrate in methanol). A spill that came into contact with cotton clothes would burst into flames. To this day, the *Komet* remains the only rocket powered fighter.

Fig. 2 V-2 (1944)

49

Sputnik opens the Space Age

After World War 2 Russia and America each grabbed every German rocket scientist they could find. Russia's German scientists were faster than our German scientists and they opened the Space Age for real with **Sputnik 1** (Fig. 3)**,** the first man-made satellite to orbit Earth in 1957. To a 12 year old American boy growing up in Ireland, it was embarrassing. Every time it passed overhead we could dial its frequency and hear Beep – Beep – Beep.

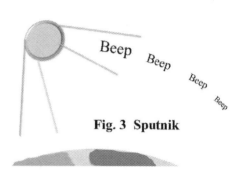

Fig. 3 Sputnik

There had been no announcement prior to the launch, and one morning we woke up to hear that the Space Age had begun for real. A few months later, with great fanfare and the eyes of the world watching, America's first attempt blew up on the launch pad! Russia's success continued with a dog, Laika, and then the 1st man in Space, Yuri Gagarin in **Vostok 1** in 1961.

America's pride was hurt, so money poured into STEM education. I was one of the lucky recipients of this government largess and rode it all the way to a PhD in Chemistry. What does that have to do with the Space race you might ask? Very little really, even though my last three years in graduate school were on a NASA Traineeship. Chemistry was 'good', as was science in general. As a result, the USA beat Russia to the Moon in 1969. No other country has been able to put men on the Moon in the 40 plus years since America's Apollo program.

What happened since those glory days of scientific advancement? Some chemical companies removed the word "chemical" from their corporate names. DuPont used to advertise "Better living through chemistry", and no longer does so. Something happened to make chemicals evil. Napalm, nausea gas, and Agent Orange in the Viet Nam War, and widespread opposition to the war were a major factor. Global warming hysteria subsequently made technology something to be feared by the ignorant masses.

NASA retired its Space Shuttle fleet without having a replacement for it, so America now has to rent space on Russian rockets to get to and from the International Space Station. Fortunately, Yankee ingenuity has survived and private corporations now compete for Space tourism. Reservations can be made to fly on **Virgin Galactic** (www.virgingalactic.com) which will be launched from its mother-ship **White Knight Two**. On the other hand, American students now rank only 25th out of the top 29 countries in Math, and are similarly deficient in the sciences. The Digital Age is not a time to be behind in these subjects. We need another **Sputnik** shock!

Virgin Galactic emerged from the design shop headed by Burt Rutan, the designer of the **Varieze**. Burt Rutan is a perfect example of a paradigm shifter.

Chapter 11 Pioneers Today –Visionaries & Loons

A paradigm is a rule that is followed, often without knowing it. An example is the design of the bicycle saddle. It is called a saddle rather than a seat, because bicycles replaced horses. But there is no horse and no need to challenge the male anatomy so cruelly. Paradigm paralysis is why many breakthroughs come from outside the field. Paradigm shifters are either outsiders who don't know better, or insiders who deliberately identify and challenge existing paradigms.

The ***Wright Flyer*** reflected several paradigms. Birds turn by twisting their wings, so the Wright brothers used wing warping. The only propellers they ever saw were at the back of a boat, so their propellers pushed rather than pulled. Where else would they put the rudder but at the back? Because it would be pushing against air rather than water, the rudder extended farther back to gain leverage in accordance with the Principle of the Lever. This introduced a degree of flimsiness. The need to balance the rudder put the elevator equally far in front. In fact they called it a "horizontal rudder". Once people got beyond these paradigms, ailerons replaced wing warping, elevators moved to the tail, propellers moved forward, and a whole new set of paradigms became entrenched for later generations.

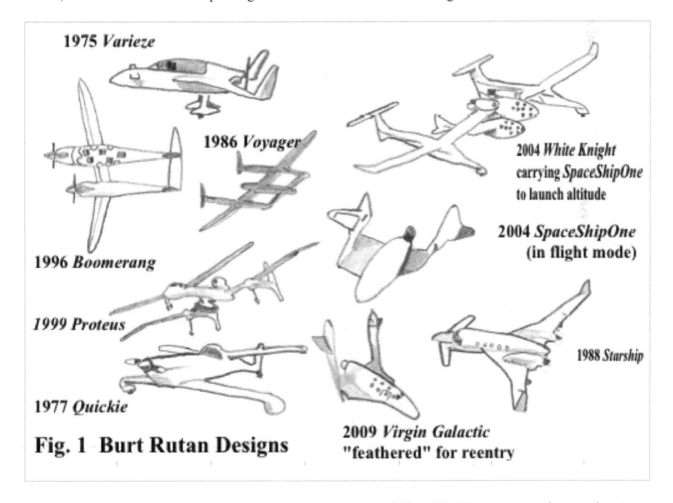

1975 *Varieze*

1986 *Voyager*

2004 *White Knight* carrying ***SpaceShipOne*** to launch altitude

2004 *SpaceShipOne* (in flight mode)

1996 *Boomerang*

1999 *Proteus*

1988 *Starship*

1977 *Quickie*

Fig. 1 Burt Rutan Designs

2009 *Virgin Galactic* "feathered" for reentry

Burt Rutan (Fig. 1) meets the definition of a paradigm shifter. ***VariEze*** was a good example. A conventional airplane's propeller pulls the plane forward by propelling air back, some of which impinges on the windscreen, wing roots, etc., effectively pushing them back. Its horizontal stabilizer is oriented so that it is being pushed down at cruise speed to ensure 'positive stability' should the engine quit. The resulting slower air over the tail would result in a safer nose down attitude. However, it also means that a 2,000 lb. plane must generate 2,200 lbs. of lift to fly. *VariEze* avoids both problems. Its propeller is at the rear, and the canard elevator in front stalls at a higher speed than the main wing. Thus, when a *VariEze* approaches stall speed, its nose drops and the main wing never stalls.

Rutan's single-seat *Quickie* echoes the look of X-Wing fighters in *Star Wars* which also came out in 1977. It is an off-set biplane with the majority of its lift generated by the forward wing, which also serves like a canard elevator.

When Burt was designing a plane that could circumnavigate Earth non-stop & un-refueled, I heard him describe how components were selected:

"We'd throw it up in the air. If it came down, it was too heavy and we'd keep looking."

Voyager had a 110.8 foot wing span and empty weight less than my Piper *Cherokee*. Fully loaded, it weighed more than 10 times its empty weight and required a 14,200 foot take-off roll. The 9-day, 28,000-mile journey used almost all of the 1,489 gallons of fuel.

Proteus is a telecommunications platform that can be flown with or without pilots on board. It was designed as an alternative to costly satellites and can fly at >60,000 feet for 18 hours. It would be far less expensive to have three *Proteus* aircraft than a single satellite. That would allow a city or an army to have one aloft at all times to broadcast TV signals. When something needed to be changed or repurposed, it would be a lot easier to do with Proteus than with a satellite. It is currently owned by Northrup Grumman and is being marketed as a multi-mission vehicle.

Beechcraft Starship was designed by Burt Rutan's Scaled Composites and constructed by the Beech Aircraft Corporation. It was intended to be a 6-8 seat business transport, but is most famous for being the chase plane in test flights of SpaceShipOne.

Recognizing that rockets are at their most dangerous close to the ground, *SpaceShipOne* and *Virgin Galactic* avoid the low altitude hazard by being carried aloft to safer ignition altitudes by their *White Knight* mother ships. People are already making reservations to fly into Space on Virgin Galactic.

Finally, *Boomerang* ignored many design conventions to minimize the asymmetric thrust of a twin engine airplane should one engine quit. Its forward canted wings have different lengths and the right wing extends forward from the fuselage at a higher angle than that of the left wing. One engine is at the nose of the fuselage and the other is on a boom like an outrigger that joins the left wing to the outer end of the empennage that (surprisingly for a Rutan design) is at the back. The engines are close enough to each other, so that asymmetric thrust is not a problem during single engine operations.

If you looked into the aircraft when it was first displayed at the Oshkosh Fly-In, you might have missed the instrument panel. When it was in the aircraft, it consisted of a Mac Powerbook which was hooked to both the engine and a GPS. It displayed both the current engine condition and its history in a very innovative format. Burt's description of how the signal from a GPS, plus a program, can replace panel gauges such as turn & bank, altimeter, VSI, airspeed etc. was a real eye opener.

Burt Rutan now has a website (http://burtrutan.com) that says more than I can say, so I will let you read more about this remarkable designer there. One final point that I will make relates to a talk Burt gave during the Oshkosh Fly In that caught many people by surprise. Burt is a committed skeptic about anthropogenic global warming. When I spoke to him afterwards and shared a presentation I had made on the subject, he honored me by having my talk posted alongside his on the website.

Wing Suits   (http://www.metacafe.com/watch/1504805/wingsuit_base_jumping/)

Click on the above link and you will see grown men playing out a fantasy (Fig. 2) they have had since watching Rocky & Bullwinkle in Saturday morning cartoons. Rocky was a flying squirrel who wore a leather helmet & goggles.

Wing-suit flying is an off-shoot of BASE jumping. That stands for Buildings, Antennas, Spans (i.e. bridges), and Earth (cliffs). BASE jumpers wear parachutes that they deploy before hitting the ground. Aside from the hazard of the landing itself, BASE jumpers incur the risk of impacting other buildings, wires, or the cliff face, both before and after deploying the canopy.

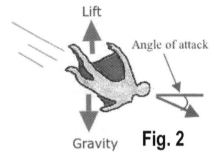

Fig. 2

The main difference between BASE jumping and delayed release sky-diving is proximity to the surface and the inability to deploy a back-up parachute should something go wrong with the primary chute. Wing-suit flying is a different kettle of fish. In addition to a parachute (for which there is no back-up), the wing-suit flyer has three wings that prolong the fight. One is between the legs and two are from wrists to the nearest thigh. Gravity provides thrust and the positive angle between the wings and direction of flight provides the lift. Although descending, relative to the oncoming wind, the leading edge is higher than the trailing edge.

Wing-suit flying is as close as man can get to flying like Superman. The people in the above video fly at 90 mph just inches from the cliff face before turning out into a valley and deploying their parachutes. While undoubtedly they would like to be able to land without using a parachute, the physics of wing loading just will not allow it.

Wing loading, the ratio of weight to wing surface area, determines the velocity needed to provide lift for flight. With a wing loading of 90 pounds per square foot (psf), a wing-suit needs a velocity of 90 mph according to an article by William Speed Weed in the July 2003 issue of *Popular Science*. For comparison, the wing loading on a Cessna 172 is 14 psf and it can safely fly at 60 mph. A flying squirrel has a wing loading of 1 psf and can fly at 3 mph. There is no need for a parachute at that speed. Undoubtedly, more wing area would reduce the wing suit's required air speed, but the increased force might rip the flyer's arms off, particularly if the increase was achieved by extensions to the arms. If you want bigger wings, get a hang glider. It would be braced for it.

Wing-suits are more complicated than they appear at first glance, and early generations reportedly killed 72 out of 75 pioneers. The ***Bird-Man*** wing-suit (Fig. 3) has three wings that self-inflate through mesh air inlets under the armpits and crotch. Air that accumulates inside the wings adds camber and stiffness to the wings, similarly to ram-air rectangular parachutes.

A conventional parachute harness would interfere with the leg wing, so the leg straps for the parachute on a wing-suit wrap around the thigh inside the suit. Zippers at each wrist frees the arms from the wings when one needs to deploy the parachute. Otherwise the flyer could not raise his arms overhead to control it. Without control one might wind up landing on a tree or on top of an airplane, and either event could be life

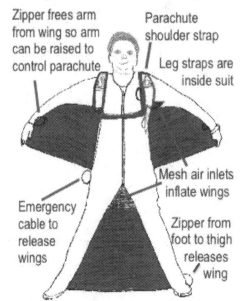

Fig. 3 Wing-Suit (see www.bird-man.com)

threatening. Many pilots regard sky-divers as akin to hooligans and might react violently to somebody landing on top of their airplane. Zippers from foot to thigh free the legs to allow the flyer to run during landing.

Who has not experienced a stuck zipper, so there is redundancy in case either zipper sticks. Emergency handles at the hips are attached to cables that run up the sides of the flyer's body. If you pull them out, the wings detach.

Today's Pioneers of Aviation

Although man has been flying consistently since 1783, there are still plenty of pioneers willing to take to the air in what most people would regard as 'high risk' aircraft. Are those wing-suit guys a bunch of loons? (I think so) But the reality is that this was the case for many of the early aviators. How about Jonathan Trappe's unsuccessful attempt to fly across the Atlantic in 2013? His "aircraft" was a cluster balloon, i.e. a chair tied to 365 helium filled balloons. 150 years after Solomon Andrews' *Aereon 1*, this was actually a step back in time. *Aereon I* in 1864 was also a cluster balloon, in that the three long cylinders each had seven smaller balloons inside them. However, Trappe's cluster balloon is far more colorful. (see http://clusterballoon.com/)

"Jet Man" Yves Rossi is often referred to as "Rocket Man". He has flown across the English Channel and given exhibition flights in many parts of the world. Although he "wears" the aircraft and lands by parachute just as Wing Suit flyers do, he is definitely a step beyond them. He straps himself to a rigid wing that has four jet engines attached underneath. The four jet engines caused him to run afoul of bureaucracy when an FAA official decided that a multi-engine type rating was required. You can read more about him at: (http://www.jet-man.com/prod/index_en.html/)

The *Terrafugia* flying car (http://www.terrafugia.com/aircraft) was demonstrated during the Oshkosh Fly-In in 2013. The wings fold up when the car is to be driven on the road. Although it may be street legal and able to fit in a 1-car garage, this flying machine will be pricy. As stated earlier, the requirements for an innovation to be considered "good" are that it be both new and reproducible. Another requirement needs to be added. It must also be affordable.

While the cost of these innovations range from a few hundred dollars (cluster balloon) to a few hundred thousand dollars *(Terrafugia)*, to millions *(Virgin Galactic)*, their initial usefulness will be for excitement and fun. However, that was also true of balloons in the 19th century, and look at where they drove aviation. Once the flight Genii was out of the bottle, it could never be put back.

Chapter 12 STEM Q&A – Aviation History

Cover the answers while trying to answer the questions.

Questions	Answers
Who piloted 1st successful* manned flight? (a) Orville and/or Wilbur Wright (b) Abas Ibn Firnas (c) Pilatre de Roziere and/or Marquiz d'Arlandes (d) Joseph and/or Etienne Montgolfier * "successful" = You walk away & can do it again	**(c) Pilatre de Roziere and/or Marquiz d'Arlandes** They flew the Montgolfier Hot Air balloon. The Montgolfier brothers appeared to be chicken and petitioned the King for the "loan" of two condemned criminals. As Pilatre had already gone aloft on a tethered flight, he objected and became the 1st aeronaut along with Marquiz d'Arlande.
Who was the first to fly across the Atlantic? (a) Charles Lindberg in 1929 (b) John Alcock and Arthur Brown in 1919 (c) Thaddeus Lowe in 1865 (d) Douglas "Wrong Way" Corrigan in 1938	**(b) John Alcock & Arthur Brown in 1919** They flew a Vickers Vimy bomber from Newfoundland and crash landed on a bog near Galway in Ireland. The bog looked like a smooth field, but was too soft for a landing.
What was the largest aircraft in history? (a) Lockheed Galaxy C5A (b) Airbus 820 (c) Hindenburg (d) Boeing B-52 Strato Fortress	**(c) Hindenburg** At 803 feet long and 135 ft in diameter, it is still the largest aircraft to fly.
Who is the 1st known survivor of rocket powered flight? (a) John Glenn 1962 (b) Yuri Gagarin 1961 (c) Alan Shepard 1961 (d) Lagari Hasan Celebi 1633	**(d) Lagari Hasan Celebi** In 1633 he attached 7 rockets to a barrel. He launched near the Bosphorus (Turkey), and descended by parachute into the water. He never flew again.
Who might English children claim was the first man to fly? (a) George Cayley's coachman (b) George Cayley (c) Robert Louis Stephenson (d) Jules Verne	**(a) George Cayley's coachman** By 1853 George was too old (85) to be a test pilot, so he got his coachman to do it. The coachman quit immediately afterwards. Today, nobody knows his name.

Questions	Answers
Which law(s) explains the LIFT of 1st manned aircraft? (a) Newton's 3rd Law & Lever Principle (b) Bernouilli's Law & Coanda Effect (c) Charles' Law & Archimedes' Law (d) Avogadro's Law & Boyle's Law	**(c) Charles' and Archimedes' Laws** The Montgolfier Hot Air Balloon used both. *At constant pressure, the volume of an Ideal Gas is directly proportional to its absolute temperature* and *Any object floating in a fluid is buoyed up by a force equal to the weight of the displaced fluid*
Who was the first to fly across the English Channel? (a) Louis Bleriot in 1910 (b) Santos Dumont in 1911 (c) John Alcock & Arthur Brown in 1919 (d) Jean-Pierre Blanchard & John Jeffries in 1785	**(d) Jean-Pierre Blanchard & John Jeffries in 1785** It was a close thing. Having to lighten the balloon they threw everything that wasn't necessary overboard and arrived in their underwear.
Which one flew before the other? (a) Wright Flyer (b) Zeppelin	**(b) Zeppelin** The first Zeppelin flew in 1900 and was bigger than a B-747. Zeppelins were the 1st revenue generating airliners and the 1st trans-Atlantic airliners
Who piloted the 1st recorded Moon shot? When? (a) Jules Verne 1898 (b) Yuri Gagarin 1959 (c) Neil Armstrong in 1969 (d) Wan Hu in 1500	**(d) Wan Hu in 1500 AD** He must have been a bit of a nut job. He attached 47 rockets to a sturdy chair. (He probably did not get to the Moon, but who knows?)
Who do Arab children say was the 1st man to fly? (a) Lagari Hasan Celebi (b) Abdul Abulbul Emir (c) Abbas Ibn Firnas (d) Osama Bin Laden	**(c) Abbas Ibn Firnas** In 875 AD he flew off a Minaret in Cordoba, Spain, using a hang-glider. The flight went well until he landed. He was carried away and never flew again. Thus, it does not qualify as a "successful" flight. Baghdad International Airport is named after him

Chapter 13   # The M in STEM

The 'M' in STEM stands for math. To many people math is just addition, subtraction, multiplication, and balancing a checkbook. To navigators it is all those plus applied geometry.   For bankers, math includes interest rates and mortgage tables. Engineers use calculus. Physicists use equations replete with upper and lower case Greek letters. (Dating sorority girls in college taught me the upper case Greek alphabet. I hate lower case Greek letters!)

At the end of the day, math is just another language used to communicate specialized knowledge. If you can learn French or Latin grammar, you can learn math. However, unlike a language, you can use a math equation without being able to derive it in the first place.

I will try to demonstrate that in the next few chapters, starting with an equation you have already seen, the Ideal Gas Law. Then I will show how simple high-school geometry opened the door for some of the greatest discoveries, specifically, the world in which we live. I will conclude with an equation that can only be described as a humdinger, the Orchard Equation.

$$PV = nRT$$

Where P = Pressure   n = Number of moles of gas   T = Absolute Temperature
       V = Volume   R = a numerical constant

Although it may not look like math, the Ideal Gas Law is math. It is also a "unifying equation", one that pulls all the other gas laws (Boyle's, Charles', and Avogadro's) together to describe the relationships between temperature, pressure, volume and the number of gas molecules. If you want to increase the pressure (P) in a given amount of gas (as in a steam engine), the equation tells you that you either have to increase temperature (T) or reduce volume (V), or both. In this example n is fixed and is the given amount or moles of gas. The Ideal Gas Constant – R – is a constant number by definition, and it never changes as long as you are using the same units. Another example is an around-the-world flight by a Roziere, a combination hot air and helium balloon. You know that the gas will cool at night, so the gas envelope will shrink and displace less air. You can compensate by heating either the helium or the air inside.

Expanding STEM is usually a stepwise process

Progress is like adding ten more miles to 100 miles of railroad. To turn a 100 mile length of railway into 110 miles, one does not have to rebuild the 100 that already exist. Science works similarly, as long as the next scientist can understand what the earlier scientist determined. That is where math comes into play. Scientific progress is like adding miles to that railroad. Boyle's Law was followed by Charles' Law, which was followed by Avogadro's Law. Each of those scientists built on the work of earlier scientists and the result was the Ideal Gas Law. Byproducts included the steam engine, the Industrial Revolution, and balloons that could fly around the world.

Navigation is a form of "applied geometry", and geometry is itself a form of math. The measurement of latitude is surprisingly simple, whereas determining longitude was relatively complex until watches and clocks capable of functioning during a long sea voyage became available.

Measuring Latitude Uses Simple High School Geometry

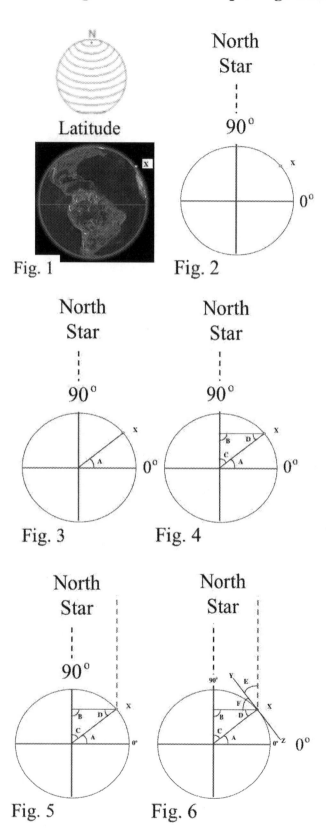

North Star

90°

Fig. 1

Fig. 2

North Star

90°

Fig. 3

North Star

90°

Fig. 4

North Star

90°

Fig. 5

North Star

90°

Fig. 6

As will be discussed in later chapters, the north pole points almost directly at the star Polaris (a.k.a. the North Star) at all times of the year. If you can measure the angle between the horizon and Polaris you have a direct determination of your latitude. X marks the spot (Fig. 1), and you want to determine its latitude. The geometry is simple:

Fig. 2 shows a cross-section of Earth. The two ends of the horizontal line through the center are on the equator at 0°. The vertical line is the axis of rotation, with the North Pole at 90°N. The dashed line into Space and is the North Celestial Pole.

Fig. 3 – Draw a line from X to the center of the Earth. Angle A is a measure of the latitude of X.

Fig. 4 – Draw a line from X perpendicular to the axis of rotation. This provides 3 new angles, B, C, and D. Because the sum of the three angles in a triangle must equal 180°, and since B = 90°, then C + D = 90°. However, since C + A = 90°, then A = D.

Fig. 5 – Draw the dashed line from X to Polaris. This line will be parallel to the Celestial Pole. You wonder how the two lines can be parallel, since they intersect at Polaris. They are effectively parallel because (a) Polaris is vastly bigger than Earth, (b) the extreme distance, 432 light years to be precise. For the Trekies, it would take 1.9 years to get there at Warp 6.

Fig. 6 – At point X, draw the line YZ perpendicular to the line from the center of the Earth, yielding two more angles E and F. Angle E is the angle of Polaris relative to the horizon. This is something that can be measured. Does it tell us the latitude of X?

Because YZ is perpendicular to the line from Earth's center, then angles D + F = 90°. Likewise F + E = 90°, so D = E. In Fig. 4 we saw that A (latitude of X) equals D, so A = E.

While measuring angle A would be impossible, measuring angle E is easy to do with tools that will be covered later, including the Cross-staff & Celtic Cross from Egypt, Astrolabe & Quadrant from medieval Europe, and yesterday's (Sextant). Today, GPS navigation is cheap and accurate, but the old ways still work.

A point I want to make in this chapter is that math does not have to be complicated to be useful in the STEM continuum. Of course, it can be complicated, but if it can be reduced to an equation whereby one can understand the relationships between different parameters – as in the Ideal Gas Equation – then you can use it without necessarily doing the calculations. Orchard's Equation is an example of this.

Orchard's Equation

This unifying equation is so intimidating that it can shut people up when you write it on a blackboard. Sometimes that is a good thing, other times not. Like the Ideal Gas Law (which now looks simple to you, right?), Orchard's Equation describes the relationship between different parameters of a liquid film, and was originally developed for paint.

Before you panic, let me reassure you that I am not going to take you through the calculation. I just want to demonstrate how seeing the relationships between parameters in an equation can help a formulator. Note how some of the parameters are to the power of 3 or 4, which makes them very powerful.

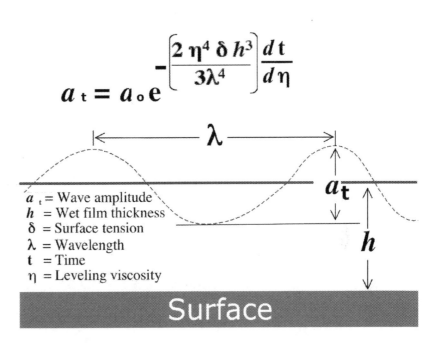

$$a_t = a_o e^{-\left[\frac{2\eta^4 \delta h^3}{3\lambda^4}\right]\frac{dt}{d\eta}}$$

a_t = Wave amplitude
h = Wet film thickness
δ = Surface tension
λ = Wavelength
t = Time
η = Leveling viscosity

Another reason is to demonstrate that complex systems usually involve compromises between many attributes, some of which are dependent on each other, and others which are independent. While the Orchard Equation relates to the leveling properties of paint and floor polishes, similarly complex equations can be written for investment strategies, mortgages, aircraft lift, etc. The breakthroughs (and big money?) go to the person who can understand and master such complexities.

For example, the wavy line in the figure is the surface pattern of a liquid paint or floor polish while it is still wet. The amplitude of the wave in wet paint immediately after application is a_o and the amplitude at some later time is a_t. To have the smoothest, flattest film, you want the amplitude a_t to be as low as possible, and the wavelength λ to be likewise. Accordingly you want everything above the line in the equation to be maximized and everything below the line to be minimized. That is what I meant when I mentioned the value of understanding the relationships between parameters, but how does the formulator use this information?

The thicker (h) the wet film, the better the leveling, so you want the liquid to be more viscous. Because you want the wavelength λ to be as short as possible, a paint brush will work better than a sponge as an applicator. You want the surface tension δ to be high to pull down the top of the wave and pull up the valleys, but surface tension cannot be so high that the paint will not wet the surface. Mercury is a good example of a liquid having a surface tension so high that it will not "wet out" on any surface. Bottom line, surface tension is an attribute where compromise is required. The dimension of time t is also important. The leveling process has to be completed before the drying process solidifies the film, and any chemical you add to delay drying has an impact on a_t, λ, and δ. I could go on, but I hope you get the point that mastering complexity is an opportunity for the person who can achieve it. Like it or not, complexity will decide who wins and who loses in the Digital Age.

Confession Time

Speaking of mastering complexity, although I used Orchard's Equation many times when I was a development chemist working on floor polishes, I never once ran the calculation. Simply knowing the relationships between the parameters told me what I needed to do. In addition, I could write it on the blackboard and people would immediately assume that I knew more than them about the topic. Nobody ever challenged it!

Chapter 14   Airplane Math - Weight & Balance

"Take them for a ride before you leave, Seán." Unfortunately, I agreed.

My first cross-country flight with family involved my 7 and 8 year old daughters. I had done a thorough preflight inspection, including calculating the weight & balance of the Cherokee. With their light weights and my own not so light, 50 gallons of fuel were not a problem even though it was a hot day. 30 minutes later we arrived at our destination having burned at most 4 gallons of gas. We met the friends with whom my girls were to stay. They had brought along some neighbors and asked me to take them up.

Three adults climbed into the plane with me … and the 46 gallons of fuel still in the tanks. It was not until I was approaching the end of the grass runway that our dilemma dawned on me. The Cherokee was slow to lift off and just as slow to climb. The nose did not want to come up, and I was not about to force it until I had more than 72 mph airspeed. That is 1.3 times stall speed and it was slow in coming. I made gradual turns towards downwind, base, and final. The flight was less than 10 minutes and we landed safely. I did not tell them that we were both over weight and forward of the allowed c.g. envelope. It was a memorable learning experience.

Unlike a see-saw, an airplane does not have a fixed point of rotation. The c.g. varies with passenger and fuel loads and it is the pilot's responsibility to make sure that the airplane is safe to fly, even if it involves asking embarrassing questions. ("Madam, how heavy are you?")

An airplane's Pilot's Operating Handbook (POH) specifies a Maximum Gross Weight and a Safety Envelope for that aircraft. A Cherokee PA28-140's safety envelope is 10 inches at 1,650 lbs. and only 4 inches at its maximum weight of 2,150 lbs.

If the c.g. is forward of the safety envelope then the plane may be so nose-heavy that you cannot get enough air passing over the elevator to raise the nose during takeoff. If it is tail-heavy, then the aircraft may stall, which can also be fatal if it occurs at the wrong time. Just because a plane has four seats and tanks for 50 gallons does not mean that you can fill them all at the same time.

The POH lists: (a) **Arm** and **weight** of empty airplane. The arm is the c.g. of the empty airplane.
                 (b) The arm of each load item, e.g. oil, fuel, front & back seats, and baggage.

Moments of each of these loads are determined by multiplying their Arms by their Weights, and the total is entered in a table. Quarts of oil and gallons of gasoline are converted to lbs. by multiplying by 1.8 and 6 respectively. The c.g. is calculated by dividing the sum of all the Moments by the sum of all of the weights, including the aircraft itself. The pilot compares the Moment and Total Weight of the loaded aircraft with the allowed limits in the POH. If the plane is over-weight or out-of-balance, then something or some one has to be moved or left out.

How Airlines Calculate Weight & Balance

Airlines rarely ask passengers for their weights. While some aircraft have weight sensors on their under-carriages, on others an average weight of 170 lbs. is assumed for passengers. That works as long as you have average passengers, but what if a professional football team boards?

One famous accident involved a small commuter airplane carrying 18 burly golfers each weighing 200 lbs. or more, and golf clubs in the baggage compartment. The plane took off and immediately assumed a nose up attitude. The pilots tried unsuccessfully to lower the nose, and the airplane stalled and crashed into a hangar.

How owners of Piper Cherokees Calculate Weight & Balance

The following chart provides an example for my own airplane. The goal is to ensure that the c,.g. of the loaded aircraft falls within the safe envelope, which extends from 84" aft of the nose to 94". Significantly, the forward limit moves back as total weight goes up.

Digital Age Tools Make Life Easier

As can be seen in the graph, the safety envelope for N4500R's center of gravity is only 4 inches wide (i.e. 90 - 94" aft of the nose) when the airplane is at its maximum gross weight of 2,150 lbs.

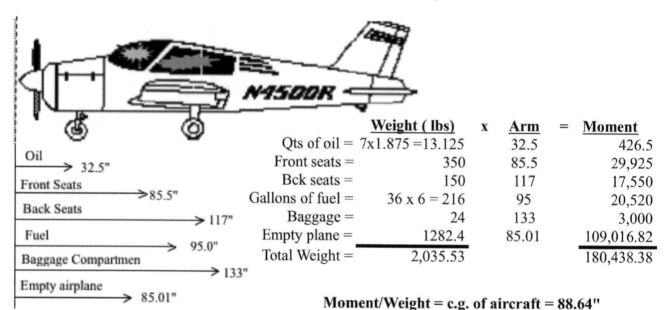

	Weight (lbs)	x	Arm	=	Moment
Qts of oil =	7x1.875 =13.125		32.5		426.5
Front seats =	350		85.5		29,925
Bck seats =	150		117		17,550
Gallons of fuel =	36 x 6 = 216		95		20,520
Baggage =	24		133		3,000
Empty plane =	1282.4		85.01		109,016.82
Total Weight =	2,035.53				180,438.38

Oil → 32.5"
Front Seats → 85.5"
Back Seats → 117"
Fuel → 95.0"
Baggage Compartmen → 133"
Empty airplane → 85.01"

Moment/Weight = c.g. of aircraft = 88.64"

As the years passed I put on more weight than is good for me or for my Cherokee. The net result is that I never fill the tanks to the full 50 gallons any more, in anticipation that somebody of equal weight might accompany me in the right front seat on the next flight. However, the calculation shown above assumes that both front seats are at an arm of 85.5", which is the 5th of 9 notches on the seat rail. Moving the right seat back provides 'wiggle room'.

To make life easier I wrote a computer program for my i-Pad which assumes that the right front seat is positioned at the ninth notch.

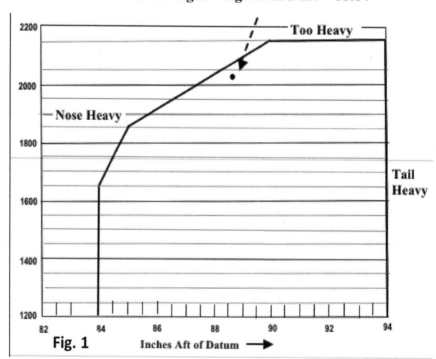

Fig. 1   Inches Aft of Datum ➝

My own seat is always in the same position. I know from experience that the maximum weight I can have in the right seat is 205 lbs. if I have 36 gallons of fuel on board (which is why I rarely fill it above 34 gallons after a flight). Some of my friends have also put on too much weight. The program follows:

Cherokee Notches

Cherokee Notches is a spreadsheet program that works on an i-Pad or i-Phone using the Numbers app. Rather than combining the weights in the two front seats together in the Weight & Balance calculation, it assumes that left front seat is in the mid position (5th notch) and the right front seat is at the aft-most notch (9th notch). Special Air Worthiness Information Bulletin No. ACE-97-02 provided the locations aft of datum for each notch in the seat rail. Whereas the nominal front seat location is at 85.5" aft of datum, this varies from 80.5" at the most forward position to 90.5" at the most rearward. The arm moves aft 1.25" for each of the 9 notches. When the front seats are occupied by two people that each weigh over 210 lbs, the ninth notch becomes very useful.

The program is quite simple. One enters the data for the flight into the 6 red boxes (B2-B7), and the program spits out either "Weight is OK!" or "Over Max Weight" in Cell C9. Where balance is concerned, Cell C10 will say either "CG is OK" or "Unsafe CG".

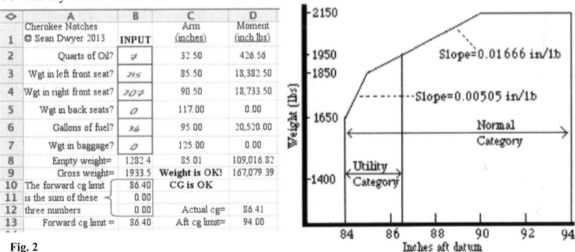

◇	A	B	C	D
1	Cherokee Notches © Sean Dwyer 2013	INPUT	Arm (inches)	Moment (inch lbs)
2	Quarts of Oil?	7	32.50	426.56
3	Wgt in left front seat?	215	85.50	18,382.50
4	Wgt in right front seat?	207	90.50	18,733.50
5	Wgt in back seats?	0	117.00	0.00
6	Gallons of fuel?	36	95.00	20,520.00
7	Wgt in baggage?	0	125.00	0.00
8	Empty weight=	1282.4	85.01	109,016.82
9	Gross weight=	1933.5	**Weight is OK!**	167,079.39
10	The forward cg limt	86.40	CG is OK	
11	is the sum of these	0.00		
12	three numbers	0.00	Actual cg=	86.41
13	Forward cg limt =	86.40	Aft cg limt=	94.00

Fig. 2

Formulas in different cells are shown below:
Enter cells in column A as shown. B8 is the empty weight for N4500R. Cell B9 provides the gross weight by converting quarts and gallons to pounds and summing weights in Cells B2-B8.
i.e.   B9 = 1.875*B2 + B3 + B4 + B5 + 6*B6 + B7 + B8

Because the forward c.g. limit (B13) changes with gross weight in three steps, it is calculated by adding the three possible solutions B10-12. Two will always equal zero.

B10 =IF(B9>1850,(85.01+0.01666*(B9-1850)),0)
B11 =IF(AND(B9>1650,B9<1851),84+0.00505*(B9-1650),0)
B12 =IF(B9<1651,84,0)
B13 =SUM(B10:B12)

Cell C9 communicates status of weight in English:
C9 =IF(B9<2150.5," Weight is OK!","Over Max Wgt!")

Cell C10 does likewise for Balance:
C10 = IF(AND(D9/B9>B13,D9/B9<D13),"CG ok","Unsafe CG")

Moments in column D are calculated by multiplying Column B (weights) by Column C (arms), while converting quarts and gallons to lbs in cells D2 and D6.   Total Moment is D9
.

D2 =1.875*B2*C2	D5 =B5*C5	D8 =B8*C8
D3 =B3*C3	D6 = 6*B6*C6	D9=SUM(D2:D8)
D4 =B4*C4	D7 =B7*C7	D12= D9/B9

Cell D13 has the Aft cg limit, which is always 94.0.

Nose Heavy/Tail Heavy – Cause and Effect

While the causes are obvious - i.e. the aircraft is loaded so as to be out of balance - what is the effect of being nose heavy Vs. tail heavy? In the case of nose heavy, the only way to get the nose to come up is to go faster and get enough air blowing over the elevator at the tail. While easy to achieve during flight, this may not the case during takeoff. It takes time to accelerate and you may reach the end of a short runway before sufficient speed is achieved.   Complicating factors are (a) gross weight, (b) temperature, and (c) the possibility of taking off into 'ground effect'.

As can be seen in the diagram above, the forward limit of the c.g. moves back with increasing gross weight. The total range of the safety envelope is only 4" at max gross weight. Temperature is a problem because higher temperature means lower air density, and lower air density requires higher speed to get enough molecules of air over and under the wings and control surfaces. Ground effect extends upwards to about half the wing span, and provides lift that is not available above that. It is possible to be able to take off and not be able to climb out of ground effect.

Tail heavy is particularly problematic after engine failure. If a plane stalls in such a situation, there is no way to get the nose pointed down so that the airplane can glide.   The plane will enter an uncontrollable flat spin. This is why airplanes are configured with 'positive stability', i.e. the tendency is to go nose down after engine failure. An airplane is controllable in a glide as long as the pilot maintains sufficient flying speed.

Most airplanes used in World War 1 had 'neutral stability'. This meant that inattention by the pilot (or incapacitation by a bullet) could result in an unexpected spin.

Chapter 15 STEM Q&A – Math

To illustrate how the four elements of STEM overlap each other, all the mathematics questions below draw on aspects of physics, chemistry, and technology that were introduced in earlier chapters. As in previous STEM Q&A chapters, cover the column on the right before answering the questions. Some will require careful thought.

Questions	Answers
Earth's gravity is 6.25 times greater than the Moon's. Using the formula (W = m x g), what is the mass on the Moon of an astronaut whose mass in Earth is 90 kg? (a) 90 kg (b) 14.4 kg (c) 563 kg	**(a) 90 kg** Mass is a constant, i.e. the amount of stuff, whereas Weight is the effect of gravity on that stuff. Thus, the mass of the astronaut would be the same on Earth as on the Moon.
Which law of science is reflected by the following math formula? **At constant pressure $V_1/T_1 = V_2/T_2$** (a) Boyle's Law (b) Charles' Law (c) Avogadro's Law (d) The Ideal Gas Law	**(b) Charles' Law** *"At constant pressure, a gas expands in direct proportion to its absolute temperature"* This explains how hot air balloons get lift. Open at the bottom, the pressure inside the balloon is equal to the pressure outside.
Which law is reflected by this formula? **M = W x A** (a) Archimedes' Law (b) Newton's 2nd Law of Motion (c) The Principle of the Lever (d) Einstein's Theory of Relativity	**(c) The Principle of the Lever** Not only is the Lever Principle used to balance airplanes, it is also used by an airplane's control surfaces (ailerons, rudder, and elevator).They use the force of air flowing past them to orient the airplane in flight.
Assuming fuel flow at 8 gal/hr, and speed =100 mph, how many gallons will you need to go 200 miles and stll have 1/2 hour of fuel in reserve? (a) 24 gallons (b) 20 gallons (c) 16 gallons	**(b) 20 gallons** That was too easy. No mention was made of wind speed or direction. In a round trip, a tailwind on one leg will be a headwind on the other. A cross-wind will work against you in both directions. A 1/2 hour reserve may not be enough!
Total Moment divided by Total Weight determines what? (a) Time to destination as a function of weight (b) Center of Gravity of an airplane (c) Maximum amount of people, bags, fuel allowed	**(b) Center of gravity of an airplane** Moment = Weight x Arm, where Arm is the distance from an some point, e.g. airplane's nose. If weight & moment of the empty plane are included, one does not need a fulcrum.
Which law is reflected by this formula? **$a = F_{net} / m$** (a) Newton's 1st Law of Motion (b) Newton's 2nd Law of Motion (c) Newton's 3rd Law of Motion (d) The Principle of the Lever	**(b) Newton's 2nd Law of Motion [$a = F_{net}/m$]** This rule reflects how a parachute works because F_{net} = Gravity - Drag. Thus, the higher the Drag of the parachute, the slower will be the acceleration. It also reflects why a golf ball hit with a golf club will accelerate faster than a bowling ball hit with a golf club. The mass of the bowling ball is so much higher than that of the golf ball.

Chapter 16 Navigation is Applied Geometry

Not only is navigation a great example of the STEM continuum, it is one of the oldest uses of STEM by man. Astronomy revealed that the North Star, Polaris is always at the same place in the night sky, and all the other stars appear to rotate around it. Some tools used to measure the angle between Polaris and the horizon were also useful for engineering and distance measurement, and math was the means by which the information was made useful. Where are we, how do we get home, how far are we off-shore? Some of the technologies were simple.

Knots or Nauts?

Speed at sea is measured in nautical miles per hour, but it is spelled "knots". Dead reckoning navigation requires awareness of the speed at which one is traveling. What did mariners use in the old sailing ships? They would throw a log overboard. It was attached to a rope that had a series of knots at specific distances along its length. The rope would be let out and they would count the knots that ran out in a certain amount of time measured by a sand clock, basically a big egg timer. They would retrieve the log and record the speed in – wait for it – a log book. That is the origin of why pilots 'log' their time as a record of flying time.

That method is also the origin of the term "knots" for the speed of a ship in nautical mph. A nautical mile, (1.151 statute miles) is defined as the distance of a minute of latitude, or $1/60^{th}$ of a degree of latitude. There are 60 minutes to a degree, so it is not metric. In later times when they knew the actual circumference of the earth, the knots on the rope were 47' 3" apart and it was used with a 30 second sand clock. If 5 knots were pulled out in the 30 seconds, then you were going 5 knots or 5 nautical miles per hour.

What is the circumference of the Earth? You just have to multiply 360° by 60 nautical miles per degree and you get 21,600 nautical miles. Multiply that by 1.151 statute miles per nautical mile and you get 24,862 statute miles for the Polar circumference. The equatorial circumference is 24,902, because the planet bulges very slightly at the equator because of the centrifugal force of rotation.

So we know how mariners measured speed, but how about direction? Lots of exploration occurred with just the stars as a guide. Knowing the shape of the Earth was a big help.

The Earth is ~~Flat~~ ~~Round~~ Almost Round

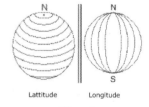

Lattitude   Longitude

Fig. 1

The key things to note when you look at a globe are:

Its **axis of rotation** is at an angle relative to the plane of its orbit around the sun, and is why we have seasons. It runs through the **north and south poles**, but not the **magnetic pole**. The magnetic north pole is not constant and is moving west at over 30 miles per year.

The **equator** is the line around the middle that is as far as you can get from the poles.

Lines of latitude (Fig 1 - left) run east-west parallel to the equator, equidistant from each other.

Lines of longitude (Fig. 1 - right) run north-south, and converge at the two poles.

The **Prime Meridian** and the **International Dateline** are longitude 0^0 and 180^0 respectively.

Tropic of Cancer is the latitude farthest north where the sun ever gets 90^0 above the horizon. This occurs at noon on June 21, the summer solstice. The northern summer solstice is also when the North Pole gets the most sunlight, at about a 66.5^0 angle. Our winter solstice is December 21.

Tropic of Capricorn is similar to the Tropic of Cancer but is south of the equator.

Arctic Circle is the farthest south latitude – in the northern hemisphere - where you get 24 hours of daylight on the summer solstice and 24 hour night on December 21. Orient a globe so that the axis of rotation is towards your face and you will be able to see everything north of the Arctic Circle and nothing south of the Antarctic Circle.

Antarctic Circle is the equivalent of the Arctic Circle south of the equator.

Analemma (Fig 2) addresses 3 issues (a) the tilted axis and (b) Earth's elliptical orbit, and (c) the Sun is not at the center of Earth's orbit. Those issues contribute to why we have **ice ages at predictable intervals**.
(That is covered in the chapter "Weather – Cause & Effect")

Analemma

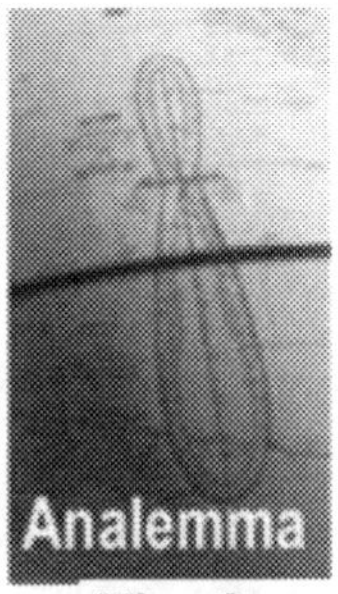

Fig. 2

Today we know that the Earth is almost round. The centrifugal force of the rotation causes it to bulge slightly at the equator. A chef twirling pizza dough demonstrates this effect. Thus, if the Earth was not rotating on its axis, the oceans would be shallower at the equator and deeper at the North Pole. The South Pole is covered by the continent of Antarctica.

While ancient farmers could be forgiven for believing that the Earth was flat, sailors who gave any thought to the matter could not. Unlike the farmer, mariners could clearly see the true horizon and would have seen ships go beyond it, with the hull disappearing first, followed by the mast. They would have known that a lookout perched in the 'crow's nest' high on the mainmast could still see the departing ship when it was no longer visible to those standing on the deck.

The only reasonable explanation was that the surface of the ocean was curved, but if the surface was curved, why did the water not flow downhill over the horizon? Figuring that out was the first step in accepting that the Earth was round. The Sun and Moon were round, why not Earth? Using the STEM concept, it was **S**cience to tell Columbus "Don't worry, you can't fall off the edge. The Earth is round". The *Nina, Pinta,* and *Santa Maria* were **T**echnology. Designing and building them was **E**ngineering. Finally, **M**ath was the language that tied astronomy to navigation and engineering to technology.

The laws of universal gravitation, published by Newton in 1687, state that the gravitational force of a spherical object can be treated as though all its mass is concentrated at a point in the middle of the sphere. So all of the oceans are attracted to the center of the Earth and not to some place over the horizon. It is also why the apple fell on Newton's head. It was attracted to Earth's center.

Gravity is why we do not have a 1,000 mph wind at the equator. You might expect that to be the case, as the Equator has to be rotating east at over 1,000 mph (i.e. 24,902 miles every 24 hours) and the atmosphere is not tied to Earth's surface as firmly as are continents. However, it must be tied to some degree by gravity, or the air would be sucked off by the vacuum of Space. We will discuss that further in the chapter *'Weather – Cause & Effect'*.

Measuring the Earth

Look up "circumference of the Earth" on the internet and you will find that the Greek mathematician Eratosthenes (276 BC – 195 BC), who was working as a librarian in Great Library of Alexandria in Egypt, calculated a remarkably accurate circumference for the Earth in 240 BC. He also determined that the Earth's axis of rotation is an angle of 23.5^0 with respect to the plane of its orbit around the sun.

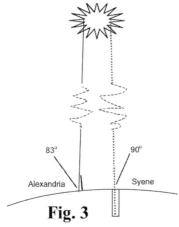

Fig. 3

Eratosthenes heard that the sun would shine down to the bottom of a well in Syene (Fig. 3) at noon on the summer solstice (June 21). That meant that the sun was directly overhead, i.e. 90^0 to the surface. Syene is close to Aswan in Egypt, which is close to the Tropic of Cancer. He then determined the angle of incidence of the sun at noon the same day in Alexandria, by measuring the shadow of a stick placed perpendicular to the ground.

Knowing the length of both the shadow and the stick, he determined that the angle of the sun was 83^0, a difference of 7^0 from Syene. Since there are 360^0 in a circle, then the distance from Syene to Alexandria is 7/360ths of the Earth's circumference. His measure of distance was how far a camel caravan could travel in one day, so it was a mite crude. However, his estimate of the circumference was within a few percent of today's accepted value.

Did Columbus know about Eratosthenes? Probably not, as Columbus underestimated Earth's circumference by about 50%. He went to his grave believing that he had discovered islands off the coast of Asia, which is why he called the islands the "West Indies" and why native Americans are called "Indians". It is also why you do not live in the 'United States of Columbia'. Amerigo Vespucci was just a few years behind Columbus and realized that there was a continent between Europe and Asia when one went west. 'America' is named after him.

History can be so cruel! On the other hand, Columbus got the credit for "discovering" land already occupied by three empires (Aztec, Inca, and Mayan) and hundreds of nations (Apache, Cherokee, Chippewa, Cree, Crow, Iroquois, Menomonee, Mohawk, Navajo, Pawnee, Seminole, Sioux, etc., not to mention the Inuit and other smaller tribes). Each had their own language, laws, and territory. Without their help, the Pilgrim Fathers would have perished in their first year. The best proof of the phrase "History is written by the victors" is the answer to the question: "Did Columbus discover America, or invade it?"

Columbus blindly followed the Trade Winds westwards in square rigged ships. They work well when the wind is at one's back, but how did he plan to find his way back to Spain, against the wind? His writings showed that he was knowledgeable about celestial navigation and that he also knew about the voyages of St. Brendan, who had made it back to Ireland. In fact, he recorded that he was looking for "Brendan's Isle" in his voyage of discovery, and would certainly have read the book about Brendan's voyages, *Navigatio Sancti Brendani Abbatis*. The book could be found in monasteries throughout Europe, and Columbus was known to have stopped in Galway, Ireland on his way back from Iceland in 1477. Nearby Galway is Clonfert, one of the monasteries founded by St. Brendan the Navigator. It certainly had a copy of the book. We will return to St. Brendan later in the chapter.

Navigation by the Stars

It is said that if your job is fun, you will never work a day in your life, and I believe that if you can't have a bit of fun in your job, then you need to get another job. In the 1980's I lived in England for about 5 years, and worked all over Europe, Africa, and as far east as India.

One night I car-pooled with coworkers to a meeting in a small hotel out in the boonies northwest of Ascot and the meeting ended after dark. We left the hotel and got completely lost. England's Motorways are like American Interstates, but it also has a zillion small roads that wind around like cow paths, intersecting with other cow paths. We were on one of them. I told the lads that I could get us back to Ascot where we left our cars if they would just stop the car and let me look at the stars. They didn't believe me, so we wandered around aimlessly for twenty minutes.

"Guys, stop the car and let me look at the sky and I'll get us home". They relented and stopped. I looked up, found the star system that we called the Plough in Ireland, and said "Go that way". In America that group of stars is known as the Big Dipper, the seven brightest stars in Ursa Major (Big Bear). They show the way to Polaris, the North Star. I knew that we were north of a major road running East-West from London. All I had to do was select whichever road went south at any intersection, and eventually we would come to it. When we did so, I did not know which way to turn to go to Ascot and just tossed a mental coin. It was either left or right, and I had a 50% chance of being correct with a guess. "Go left" I said.

We did, and eventually came to signs pointing to Ascot. Years later when I was leaving England, one of the people from that night told me that they still talked about how I had navigated by the stars. Mission accomplished! (Yankee ingenuity demonstrated)

That is how early mariners found their way home from long sea voyages from Europe to America and back to their home ports, except they would use Polaris to determine latitude as well as direction. I was only using direction, north/south/east/west when I found our way back to Ascot. That was not enough when people were navigating across thousands of miles of ocean. To get home to Europe, they would sail north or south to the latitude of their destination port, and go east along that line of latitude until they made landfall.

The first thing you need to know to use celestial navigation is that the axis of rotation of the Earth (Fig 4) points to the same place in the sky, regardless of the season. Currently, the North Star (Polaris) is within one degree of the North Celestial Pole, a line extended from the Earth's axis of rotation out into space. Its enormous distance from Earth, 430 light years, is why Polaris does not appear to move from summer to winter. A 'light year' is the distance light travels in one year. To the Star Trek generation the speed of light cubed is 'Warp 1'. It would take 1.99 years to get to Polaris at Warp 6 (i.e. 6 to the power of 3 = 216).

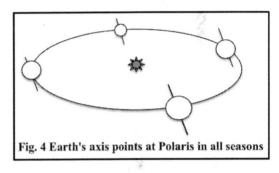

Fig. 4 Earth's axis points at Polaris in all seasons

It is easy to find the North Star when you twirl a star map around the Earth's Celestial Pole. It is the only one that does not move. Another way uses the group of stars known as the Big Dipper in the USA. It points to Polaris (Fig. 5). While the Big Dipper will appear to rotate half way around the sky in 12 hours, Polaris will appear to not move. That last sentence used the word 'appear' twice, and that is because it is the Earth that is revolving and not the stars. Your senses tell you otherwise.

As shown in Fig 5, Polaris is the first star in the Little Dipper. Because all the other stars appear to rotate around Polaris in the northern hemisphere, they can be used to tell not only time, but also longitude if one knew the local time and had the right tables. Kochab is particular was used to tell time. More about that later, determining latitude via Polaris came first.

If you are standing on the North Pole, Polaris is directly overhead, i.e. at an angle of 90° relative to the horizon. Someone on the Equator would see Polaris right on the horizon, i.e. at 0°. Simple high school geometry explains why, as described in the chapter titled "The M in STEM". Depending on the tool used to measure the angle between Polaris and the horizon, one may need to actually see the true horizon.

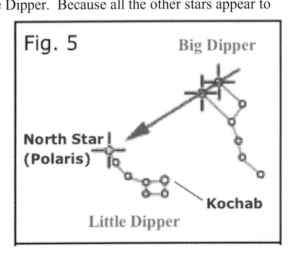

Fig. 5 Big Dipper / North Star (Polaris) / Kochab / Little Dipper

Like a pilot shooting an instrument approach to a runway, a ship sailing from New York to Galway would sail northeast until it intersected Galway's latitude, and then stay at that latitude until they made landfall in Ireland. From there on it was matter of recognizing landmarks to find the port.

Navigation Tools

Fig. 6 Sextant

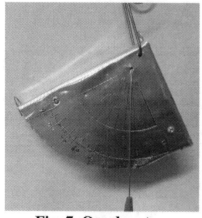

Fig. 7 Quadrant

Fig. 8 Astrolabe

What tools did mariners use? The three shown in Fig 6-8, sextant, marine quadrant, and mariner's astrolabe, are all refinements of the same idea, which is to measure the angle of a celestial body such as Polaris or the sun with respect to the horizon. Although largely displaced by GPS today, the sextant is the most modern and precise of the three. It involves two mirrors, a telescope, and a lever with an arc of 60 degrees, i.e. 1/6th of a circle. The arc and mirrors allow the navigator to see both the horizon and the sun at the same time. Because one has to be able to see the true horizon to use it, the sextant is not much use on land.

The quadrant and astrolabe both use gravity, so in effect they provide their own horizon. They were particularly reliable on land, where the problem of a rocking boat would not be faced. As its name implies, the quadrant is ¼ of a circle. One sights the celestial body through sights at the top and then reads the angle of the plumb bob. The angle of Polaris, the North Star , is a direct measure of latitude.

The astrolabe uses 1/2 of a circle. The fat part at the bottom of the cross in the center is actually a weight that serves the function of the plumb bob in the quadrant. When I saw the astrolabe, it brought to mind the Celtic Cross, which was a symbol of St. Brendan the Navigator (484-577 AD) whose voyages to Iceland and islands west of there were well known in Medieval times. In fact, many Irish people believe that Brendan sailed to America about 500 years before Leif Erickson and 1,000 years before Mediterranean Johnny-Come-Latelys like Columbus, Pizzaro, and Amerigo Vespucci.

Could Brendan have sailed to America? Yes, he could have. Tim Severin proved that in a voyage in which he used a curragh similar to the one used by Brendan. You can read about it in the book, *The Brendan Voyage*. Did Brendan get to America? That has yet to be proven. Nevertheless, the *Icelander Saga* talked about displacing Irish monks when Vikings under Eric the Red arrived in Iceland, and three of the Greenlander Sagas appear to describe meeting Irish people in lands west of Greenland.

When developing a class on navigation by the ancients for a youth program, I followed up on the possibility that there was a connection between the astrolabe and a Celtic cross. Having grown up in Ireland, I was very familiar with Celtic crosses, as they are a common gravestone (Fig. 9) and also are used on roadsides to commemorate where heroes died for Ireland (Fig. 10). As a child I believed that the circle was a pre-Christian sun god superimposed on the Christian cross by the newly converted Irish. As I found out, it may have been far more complicated. Are you ready for a little Indiana Jones?

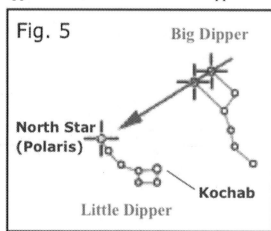

Crichton E. M. Miller, in his book *The Golden Thread of Time: A Voyage of Discovery into the Lost Knowledge of the Ancients,* claimed that the Celtic cross was originally a tool used by Egyptian engineers. He made a point about the nub at the center of the cross in the earliest Celtic Crosses, which is still seen to this day. Does it represent a nut and bolt around which the circle swiveled? His inspiration was a broken engineering tool that one of the builders of the Great Pyramid of Khufu in Egypt hid in the north shaft of the Queen's Chamber. It was found in 1872, but its significance was not understood prior to Miller. Bottom line, the working Celtic cross may have predated Christianity itself, as the pyramids were constructed several thousand years BC. Significantly, a tool that could be used to determine the height of a doorway in a pyramid could also be used to measure the angle between Polaris and the horizon.

Fig. 9 Celtic Cross

My reconstruction of the tool is shown in Fig. 11. The staff and crossbar are fixed and the circle rotates freely, although the weighted bottom - as in an astrolabe - ensures that it will always orient vertically. That feature provides a built-in 'horizon finder', which is why it could also be used as an engineering tool. One sights Polaris or a point on a pyramid along the top of the crossbar and reads the angle where the crossbar intersects the scale on the rotating circle.

Fig. 10 Roadside Memorial

Could Brendan the Navigator have known about the navigational uses of this Egyptian pyramid engineering tool? Surprisingly, although Crichton Miller did not make the connection, there is a connection. I could make it because I grew up in Tralee, Ireland. Brendan grew up and sailed from Tralee Bay. Just south of Tralee is Scota's Grave where the daughter of a Pharaoh is buried under a large stone surrounded by other large stones. Most of the people in Tralee could tell you where Scota was buried. Scotland got its name when Scota's descendants ruled that country.

Fig. 11 Navigation tool?

Sounds fantastic? Look it up. While you are at it, see if you can find any connection between the people who killed Scota (the Tuatha De Danann which is Gaelic for either "the people of Danu" or "the people of the god Danu") and Egypt. One source on the Internet speculates that the Tuatha De Danann might have been the Tribe of Dan, a lost tribe of Israel that would have been driven out of Egypt in the Exodus, which occurred approximately when Scota arrived in Ireland. Egyptologist Lorraine Evans book, *Kingdom of the Ark* describes her search for Scota, and concluded that she was Meritaten, the oldest daughter of the Pharaoh Akhenaten and Nefertiti. Tutankhamun was the son of Akhenaten by another wife, Kiya, and succeeded his father as Pharaoh.

It would be understandable if you were skeptical about this story, as are many Egyptologists. It does not fit the conventional wisdom of the field (paradigm paralysis?), in spite of a body with Egyptian jewelry being discovered during an excavation of the Mound of Hostages near Tara in Ireland (http://grandestrategy.com/2009/07/594949-story-of-princess-scota.html). Why would an Egyptian princess come to Ireland in the Bronze Age? For starters, there was a reason for why she had to get out of Egypt. Her father had introduced monotheism to Egypt, specifically Aten the Sun God. (Remember what I said about the circle in the Celtic cross?) Aten displaced the gods Amun, Horus, Isis, Ra, Osiris, and many more, and put their multitude of priests and servants out of jobs. At that time every big city in Egypt had its own god, and each god had his own cadre of supporting priests. Gods were big business. Akhenaten moved his capital out of Thebes with its Temple of Karnack and built a new city, Amarna.   Amarna was demolished after his death and none of his children survived very long after him. So it is not surprising that Meritaten, a.k.a. Scota, needed to flee.

If this Egyptian connection is of interest to you, then you might want to follow the work of Dr. Zahi Hawass on the Discovery and National Geographic channels. As the senior archeologist in the Egyptian government until the overthrow of Mubarack, he was present when any newly discovered tombs were opened. In the last few years several tombs involving Akhenaten's family have been the subject of his TV specials.

St. Brendan the Navigator

A book describing Brendan's voyages, *Navigatio Sancti Brendani Abbatis*, could be found in monasteries throughout Medieval Europe. Columbus was aware of it, and it may have been why he was so confident that he could get back to Spain in spite of the winds. That book makes fascinating but difficult reading. It described seeing floating 'Crystal Palaces'. That undoubtedly referred to icebergs, something that one never sees in Ireland, even though it is as far north as Hudson Bay and the Aleutian Islands. Ireland has a more temperate climate because of the warm Gulf Stream. (See more in Weather – Cause & Effect)

Another interesting section read as follows:
> "...they heard the sound of bellows and thud of hammer and anvil. An islander came out of a forge, caught sight of the curragh [Brendan's boat], and went back indoors. Brendan told his men to row and sail as fast as they could to try to clear the place. But even as he spoke, the islander reappeared and hurled a great lump of slag at them. It flew two hundred yards over their heads, and where it fell, the sea boiled and smoke rose up as from a furnace. When the curragh had gone about a mile clear, more islanders rushed down to the shore, and began hurling lumps of slag at the monks. It looked as if the whole island was on fire. The sea boiled; the air was filled with howling; and even when they could no longer see the island, there was a great stench. Brendan said that they had reached the edges of hell."

"The edges of hell" or an island undergoing a volcanic eruption and natives desperate to get picked up by Brendan's boat? Undoubtedly, this occurred either at the Azore Islands or Iceland, two places consistent with the description of his travels. Enough nostalgia about St. Brendan the Navigator, and it is nostalgia for me, as I grew up where Brendan grew up and died.

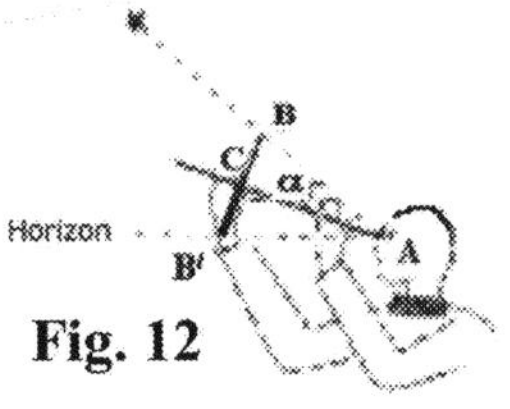

Fig. 12

Cross Staff

One other tool used to measure angles of celestial bodies that deserves mention is the astronomer's cross staff. One moves a short staff along a ruler until the star and the horizon, or a star and planet, can be seen through sights at both ends of the short staff (Fig 12). After that geometry determined the angles between the two points. It could also be used to measure distance from shore if one knew the height of a lighthouse or hill on which one was sighting.

A more modern use was in the Dam Buster's Raid in WW2 on 3 dams in Germany's Ruhr district, as described in the book *Enemy Coast Ahead* by Guy Gibson. Gibson and his squadron of Lancaster bombers had to drop a bomb from precisely 60 feet above the water at night and at an equally precise distance from the dam so that it would bounce over a torpedo net, but not over the dam.   The 60 foot

Fig. 13 Ptolomy

altitude was achieved by aligning two lights set at precise angles under the plane. Distance was achieved with what was effectively a fixed astronomer's cross staff. I'd like to say that this use of geometry and an ancient tool was classic Yankee Ingenuity, but everybody involved was from the British Commonwealth. Nevertheless, it was a great use of high school math and technology from ancient Egypt.

It is easy to confuse the astronomer's cross staff with a religious cross in old drawings. One was used by Ptolomy (90-168 AD), a mathematician, astrologer, astronomer, geographer, and poet who lived in Egypt under Roman rule (Fig. 13). He developed star charts and tables predicting the positions of stars and planets which survived into the time of scientific discovery during the Renaissance, i.e. the 14-17th centuries. There was one problem with his work: It assumed that the sun, stars, and all the planets revolved around the Earth. Galileo (1564-1642 AD) was nearly burnt at the stake for challenging this belief. Contrary to what Al Gore says regarding global warming, consensus is not science, and wasn't science back then either!

Time was key to Determining Longitude

If you know what time it is, there are several ways to determine longitude. It is also useful to know when to wake up your replacement when you are on 'late watch' on a ship, and which is why the term 'watch' was applied to 'personal clocks'.

One can determine time at night by using the stars, since they all rotate around Polaris, the North Star (Fig. 14). Columbus could tell time to within a quarter of an hour using a Kochab Clock (Fig. 15), according to *Admiral of the Ocean Seas* by Samuel E. Morison. Having tried it, I take the ¼ hour claim with a grain of salt. Nevertheless, the Kochab clock should give a rough estimate of the passage of hours at night. For example, three hours after Kochab is at the East Shoulder, it will be at the Head in the Kochab clock. More importantly, if you knew what time it was locally, e.g. midnight, and had tables of Kochab's position at different times of the year at midnight in London, you could determine how far east or west of London you were. In other words, you could determine your longitude.

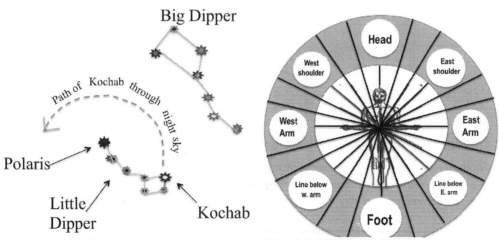

Fig. 14 Kochab circles Polaris   **Fig. 15 Kochab Clock**

Lines of longitude run north-south and converge at both the north and south poles. The entirety of each line of longitude rotates around the Earth's axis once every 24 hours, so those lines demark local time anywhere in the world and there is a direct relationship between local time and longitude. Divide 360° by

24 hours and you get 15° per hour. Accordingly, if you know both your local solar time and the current time at some other point of known longitude (e.g. the Greenwich Observatory in London), then you can calculate your longitude. Greenwich is located where longitude is defined as 0° and Greenwich Mean Time is referred to as 'Zulu Time' by pilots for weather forecasts and when filing flight plans.

Latitude could be determined by measuring the angle of the sun at noon or the angle of Polaris to the horizon, but determining longitude at sea remained elusive until people had reliable clocks so that they could compare local time with the time at any known meridian. In 1714 the King of England offered a prize of £20,000 to anyone who invented a method of finding longitude within 30 miles after a sea voyage to the West Indies. That is 1/2 of a degree. This prize was finally won with the invention of a marine chronometer (i.e. a clock) in 1761. The story is documented in the book *Longitude* by Dava Sobel, which was made into a Granada TV documentary. Their ability to get ships to rendezvous far out in the ocean, or get a fleet to a particular location on time was one of the reasons why Britannia ruled the waves for centuries.

Every day at noon a big ball slides down a pole above the Greenwich observatory. Ship captains docked in London would set their clocks as an aid for navigation after they leave. The big ball in Times Square on New Year's Eve is modeled on this.

'Apparent local time' is probably not what it says on your watch, as that is a constant across a single time zone. For example, a watch in Chicago will show the same time as a watch in Minneapolis, but the sun cannot be at its highest point simultaneously in both cities. For navigation purposes, "noon" means when the sun is at its highest point at that location. However, nothing is as simple as it seems.

Analemma

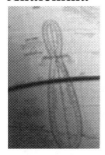

The Analemma (Fig. 16) describes deviations in the sun's position at noon in the sky throughout the year. The north-south deviations reflect the seasons and are caused by the Earth's tilted angle of rotation relative to its orbit around the sun. No surprise there.

The east west deviation is caused by the Earth's orbit, which is <u>not</u> circular. Its orbital path is actually an ellipse and it is closer to the sun in January (our winter) and farthest in July (our summer). Deviations in the ellipse over thousands of years is one of the key causes of ice ages. Another complication is that the orbital speed is faster the closer the Earth is to the sun, so sometimes the sun is ahead of where it should be and at other times it is behind. This is the reason for Analemma's asymmetrical figure 8 shape, which is smaller on top than on the bottom.

Fig. 16

The key dates for the Analemma are as follows:
April 2nd Sun's maximum offset to the east and the difference between the sun and your watch is c. 8 minutes. After that the sun appears to drift west until **July 3rd** when sun & watch time match each other. The drift west continues until maximum offset to the west is reached on Oct 2. On October 2nd the sun appears to start drifting east again until sun & watch time are matched on **Jan 2nd**.

Make Your Own Analemma

You can create your own Analemma, but it will take a year and you need to find true noon and true north at your own location. It is not enough to just look at your watch, as that will say the same time in Racine WI as it will say in La Crosse, which is on the other side of the state. It can't be noon in both places at once. To put some numbers on this:

La Crosse airport is at	N43° 52.76' /	W91° 15.40'
Racine airport is at	N42° 45.67' /	W87° 48.84'

The difference in longitude between the two cities is 3° 26.56'. When decimalized this equals 3.44° (because there are 60' per 1°). Since 15° of longitude equals 1 hour (i.e. 360° / 24 hours = 15° per hour), then 3.44° equals just under 14 minutes. That is how much later it will be noon in LaCrosse than in Racine, but what is the correction for noon in Racine Vs. Central Standard Time?

Racine's longitude puts it 5 hours and 51.24 minutes behind Greenwich Mean Time or Zulu time, so true noon will occur 8.76 minutes before my watch says it is noon, or – if my watch is set for daylight savings time then the watch will read 8.76 minutes before 1:00. Confusing, isn't it? But this is why pilots use Zulu time and the military 24 hour clock where 6:00 PM is 1800 hours. There is less confusion when everybody uses the same time system. So, knowing your exact longitude allows you to calculate the precise time when it will be noon at your location. The opposite is also true, if you know the exact Zulu time (Greenwich Mean Time) when it is noon at your location, then you can calculate your longitude.

To create your own Analemma, hammer a vertical stick into the ground in a place that gets direct sunlight around noon all year long. Then, one day every month, mark the end of the shadow at true noon by driving a small peg into the ground. Date the peg and make sure that you correct for Daylight Savings Time. The small pegs will be in the shape of the Analemma.

Compass Deviation

Does a compass always point to the north? The answer is 'No', since the north magnetic pole is not at the North Pole. In fact, it is slowing moving to the west at the moment at about 40 miles per year. This deviation is shown as "isogonic lines" on maps. When I started flying over 30 years ago, the 0° isogonic line ran north-south through Lake Michigan. In other words, 'true north' and 'magnetic north' were the same thing on that line. To determine true noon in Racine WI, all one had to do was put the 'S' of a compass next to a wooden peg and wait until the shadow of the sun fell on the 'N'. Today, however, the 4°

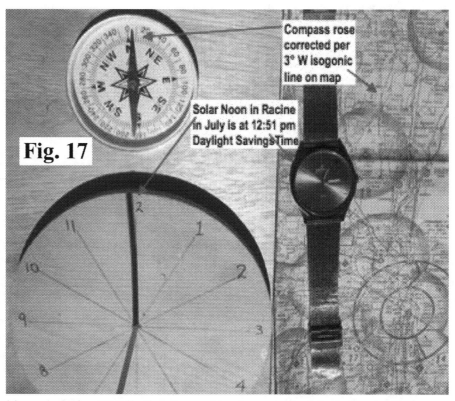

isogonic line runs down Lake Michigan and Fig. 17 tells the story. At solar noon in July in Racine WI, one's watch reads 12:51, the shadow of a peg lies at 3°

The ancients could find noon before the compass or watches were invented. All they had to do was measure the shadow of the peg. That shadow would be at its shortest at noon. However, doing this on a rolling ship would be frustrating. On the other hand, this method would not require any correction for the fact that the magnetic pole is not actually at the North Pole, so it was an excellent method when performing it on land. This method indicates True North rather than Magnetic North.

Sun Dial

Prior to the advent of railroads that needed to publish train schedules for a whole state, there was no need for time zones. Everybody operated on solar time.   In the old days you would have a sundial high up on prominent buildings in a town. Anybody could look up and read the time, as long as the sun was shining. Installing a sundial in your garden is simply a matter of sticking a peg into the middle of a circle with twelve numbers, 1 through 12, written around it, and orienting it so that 12 pointed north. The shadow of the peg would fall on the correct solar time.   One can orient the sundial so that it will read Central Standard time, Daylight Savings Time, and even put in the 8.76 minute correction for Racine.

I tried to see if I could use the sun dial technique to navigate in a plane, and it worked. We were flying back from breakfast in Palmyra and turned the plane to the south, roughly towards the sun. Then I took off my watch and turned it until the shadow of a pen placed vertically on the center of the watch lay across the watch face 1 hour and 8 minutes earlier than the time on my watch at that moment. The 6 on the watch face was then pointing due south. I looked off in the distance and found a house towards which the 6 was pointing and then turned the plane so that it was pointing at the house. Then I reset the gyro compass to match that and compared it with my magnetic compass. It was bang on, so I could lose GPS coverage AND my magnetic compass and still get direction.

A gyro compass is much more useful on a plane that the magnetic compass, because – unlike the magnetic "whiskey" compass – it does not deviate every time you change speed or make a turn. Its big deficiency is of course the fact you have to tell it where north is every 15 minutes or so.

Navigation Today – GPS

Navigation at sea is basically applied geometry with fixed reference points. Navigation on land, where landmarks such as mountains and rivers are available, is applied geometry combined with geography, unless of course it is nighttime or all the landmarks are obscured by cloud or fog. Then you would have been better off to be at sea, because you can't run into a cloud covered mountain when you are over the ocean.

Back in World War ll lost bomber pilots returning to England across the North Sea at night after bombing Germany often had to call for a "fix". This had nothing to do with the modern use of the word (i.e. "a dose of some illegal drug"). A fix to a pilot was a determination of his location by at least two, and preferably three radio stations. They would each draw a line from their location to the bomber. The lines overlapped at the bomber's position. Knowing that was the first step in finding a course home.

Celestial navigation would not have been an option when the sky was obscured by cloud in the old days. In the Digital Age that is no longer true, as the fixed reference points in the sky (i.e. satellites) are all 'visible' to the GPS receiver regardless of cloud cover, and there are so many of them that one's position can be determined to within a few feet in all three dimensions.
Nevertheless old navigation techniques still work, and pilots need to know them for that inevitable day when the cockpit electronics fail, the battery dies, etc. And that will happen.

Chapter 17

STEM Q&A – Navigation

Navigation is a combination of applied geometry, applied astronomy and geography. Don't cheat. Cover the answers while trying to answer the questions and you will learn more.

Questions	Answers
"Dead reckoning" navigation involves which of these? (a) Angle of North Star to horizon (b) Angle of Sun to horizon at noon (c) Coordinated use of speed, time, and direction (d) Blind guessing while hoping you are right	**(c) Coordinated use of speed, time, and direction** Columbus used it. So did pilots at Battle of Midway. However, he sailed at 5 knots for 3 weeks and they flew at 150 knots for 3 hours.
What symbol associated with St. Brendan the Navigator may have been both religious and a navigation tool? (a) Astronomer's cross-staff (b) Celtic Cross (c) Anchor & Fish (d) Ichtus	**(b) Celtic Cross** According to Crichton E.M. Miller's book "The Golden Thread of Time", the Celtic Cross derives from an Egyptian engineering tool. The daughter of a Pharaoh is buried near Brendan's birthplace. Working Celtic Cross
What is the single most useful star for navigators? (a) Alpha Centauri (b) Kochab (c) Sagitarius (d) Polaris	**(d) Polaris** a.k.a. North Star, Polaris is in line with the North Pole. The angle between it and the horizon indicates latitude. Big Dipper / North Star (Polaris) / Little Dipper / Kochab
Pre GPS what was the key to determining longitude? (a) Knowing both local time & Greenwich time (b) Ability to measure angle of horizon to Polaris (c) Angle of the Sun's shadow at local noon (d) Phase of the Moon	**(a) Ability to know both local time and Greenwich time** Each hour east or west of Greenwich represents $15°$ of longitude. The invention of the marine chronometer (i.e. a clock) in 1761 provided Greenwich time to ships at sea.
The Analemma is the distorted 8 found on globes. What does it represent? (a) It is an ancient Egyptian decoration (b) Corrects for Sun not being at center of Earth's orbit (c) Difference between magnetic north and true north (d) Difference between Tropics of Cancer & Capricorn	**(b) Correction for Sun not being at center of Earth's orbit** The North-South deviation is caused by Earth's tilted axis. The East-West deviation is caused by Earth's orbit being an ellipse and not circular. Also, the Sun is not in the center of Earth's orbit. Earth is closer to the Sun in the northern winter than in the summer.

Questions	Answers

Which of the United States is
     (a) Farthest south?
     (b) Farthest north?
     (c) Farthest east?
     (d) Farthest west?
	(a) Hawaii (b) Alaska (c) Alaska (d) Alaska
	
	 The Aleutian Islands are part of Alaska and they extend past the International Date Line which is both 180°E and 180°W

Why is the North Star so reliable as a navigation aid? (a) It doesn't appear to move & indicates longitude (b) It doesn't appear to move & indicates latitude (c) It is a way to tell time at night (d) It works only in Atlantic & Mediterranean, and not in Pacific	**(b) It doesn't appear to move and directly indicates latitude** Being almost directly in line with the polar axis, Polaris does not appear to move as the Earth rotates underneath. It is useful for navigation when north of the equator only, as it is below the horizon when in the Southern hemisphere. When south of the equator, navigators would use stars in the Southern Cross.

What navigation device needs a view of the horizon?
     (a) Quadrant
     (b) Sextant
     (c) Astrolabe
     (d) GPS
	(b) Sextant
	
	 The navigator measures the angle between the horizon and the Sun and then consults tables. Because you can't see the true horizon on land, it's used only at sea. 

What is this tool? (a) Mariner's Astrolabe (b) Marine Quadrant (c) Sextant (d) Compass 	**(a) Mariner's Astrolabe** It measured the angle of stars such as Polaris to the horizon, thereby showing latitude. The weight of the fat part at the bottom oriented it vertically, so a view of the horizon was not needed.
What is this tool? (a) Mariner's Astrolabe (b) Marine Quadrant (c) Sextant (d) Compass 	**(b) Marine Quadrant** It measured the angle of stars such as Polaris to the horizon, thereby showing latitude. The plumb-bob responds to gravity, thereby eliminating the need to see the horizon. Thus, like the Astrolabe and unlike the sextant, it could be used on land.

Chapter 18   Weather – Cause & Effect

This could be an extension of the last chapter.   That truism struck me when developing the navigation tools. What was useful there is also important for explaining the causes of weather. Science involves connections, and understanding one field gives you a leg up in another. You will see that explaining causes of weather is like a class in applied geography.

What causes wind? What causes currents like the Gulf Stream? About 20 years ago I saw a documentary about a survey taken on graduation day at Harvard. 23 new graduates were asked: *"What causes the seasons?"* 21 of them got the answer wrong. Such scientific illiteracy among graduates of one of our premier institutions is bad for America.

That is what got me started promoting STEM among teenagers. That survey was done in the 1980s, and by now these Harvard graduates are senators, CEOs and in other high positions. This group of scientific illiterates is running the country!

All Weather is Caused by the Sun

Without the sun, Earth would be just a barren, frozen rock in Space. Hot air, heated at the equator, rises and loses its water content as it cools. The resulting rain causes the rain forests at the Equator, in the Congo in Africa, in the Amazon jungle in South America, and in Borneo. That air is replaced by air coming from the tropics north and south of the Equator, thereby causing a circulation whereby the air that went from the tropics to the Equator is replaced from above by the now cooler and drier air that had risen from the Equator (see Fig. 1).

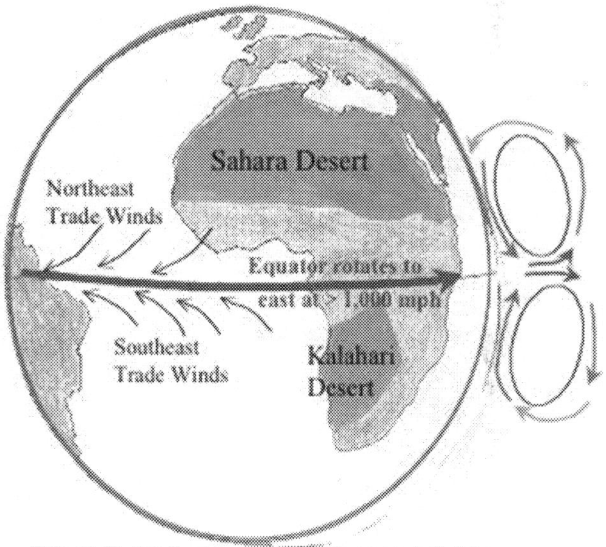

The dryness of this air is why so many deserts exist at 30° north and south of the Equator. That includes the Sahara and Kalihari deserts in Africa, the Middle East, Australia, and the deserts in Northern Mexico and south west of the USA.

Since the Equator rotates to the east at >1,000 mph (i.e. 24,800 miles in 24 hours), one might expect a 1,000 mph wind blowing from east to west. After

Fig 1 Rotation & Rising Equatorial Air cause Trade Winds, Rain Forest & Deserts

all, the atmosphere is not tied to the planet as firmly as continents. However, air is held down by gravity. Otherwise, it would have been sucked off by the vacuum of Space, as is the case with the Moon. Because air is not as firmly attached as mountains, there is some slippage. This results in a constant wind from the east at the Equator. When combined with the air coming from the tropics, this results in the NE Trade Winds north of the Equator and SE Trade Winds south of it. Pure Cause & Effect.

The effect of gravity drops off the farther one is from the surface, so the speeds of the high altitude jet-stream winds should not be surprising. However, high flying B-29s were surprised when bombing Japan in WW2 by jet-stream winds at hundreds of mph that caused them to miss their targets by miles. Knowledge of the jet-stream is fairly recent.

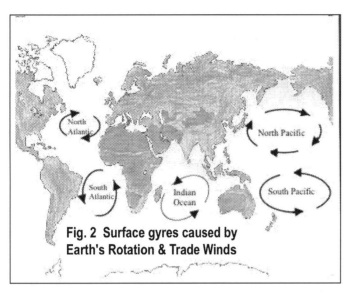

Fig. 2 Surface gyres caused by Earth's Rotation & Trade Winds

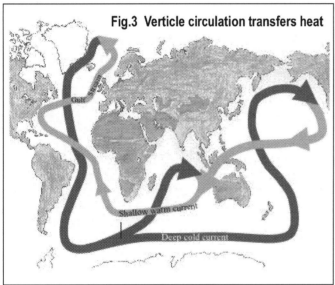

Fig.3 Verticle circulation transfers heat

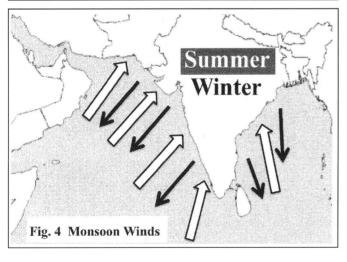

Fig. 4 Monsoon Winds

Similar circulation occurs with oceans, both horizontally and vertically. Surface water at the equator is blown westwards until it bumps into a continent. In the Northern hemisphere the deflected current rotates in a clockwise direction, whereas in the Southern hemisphere the vortex, or 'gyre', is counter clockwise, as shown in Fig. 2. One effect of this is the Gulf Stream that brings warm water on the surface from the Gulf of Mexico to Ireland and Norway. This warmth is why you can find palm trees and bamboo groves in the west of Ireland and why strawberries grow in Norway.

When the Gulf Stream has gone far enough north, the water cools and sinks into the depths and travels south. This vertical circulation is referred to as *'The Atlantic Conveyor'*. If it shut down, Europe would go into an ice age. It did shut down 13,800 years ago when the northern hemisphere was emerging from an ice age. Retreating glaciers immediately started to advance again and the ice age continued for another 2,000 years. It was why Indians never invented the wheel. If that Cause & Effect does not catch your attention, nothing will.

If the idea of opposing currents of water at different temperatures surprises you, be aware that the Germans used them to get their U-Boats past British patrols in the narrow Straits of Gibraltar during WW2. West bound warm water exiting the Mediterranean Sea through the Straits is replaced by east bound cold water at a lower depth. U-Boats would arrive at the mouth of the Mediterranean and descend until they found colder water. They would then shut down their engines and drift silently into the Mediterranean under patrolling British destroyers and corvettes. This was depicted in the German war movie *Das Boot* (i.e. "The Boat").

Monsoon Winds and Rains

The annual winds and rains associated with monsoons are both predictable and dramatic. When warm air rises in India, it cools and deposits its moisture in the form of the deluge known as the monsoon rains. In the summer hot air rising over India draws in cooler air from the Arabian Sea. These southwest winds are replaced by northeast winds in the winter when the opposite effect occurs as shown in Fig. 4. Warm air from the now hotter Arabian Sea rises and draws air from the cooling land, sweeping down from the frigid Himalayas.

These wind changes are so predictable that they have been used by traders for centuries. Sailing ships bound for India would gather in places like Somalia, Ethiopia, and the Straits of Hormuz and wait for the wind to change direction.

Historical Variation in Solar Radiation

Variation in solar activity has been recorded by court astronomers since the Middle Ages. The *Maunder Minimum* corresponded with what is known as the *Little Ice Age* (1300 – 1850 AD). The coldest point spanned the time when Washington's troops were at Valley Forge and a shortage of food triggered the French Revolution. On the other hand, the *Medieval Maximum* period of solar activity, corresponded with the *Medieval Warm Period* (900 – 1300 AD). This was when the Vikings farmed in Greenland. Today you can walk through the ruins of a farm that had stalls for >150 cattle and two other farms had stalls for 30-50 cattle. Those Viking settlements were frozen out by the *Little Ice Age*. Read Jared Diamond's book *Collapse* for more information on this.

Climate's Impact on the Availability of STEM

Science is durable, and is still useful long after the demise of the scientist. Engineering, on the other hand, requires live people who know science. Could weather be a factor in the availability of live, educated people? Could variable availability of such people be the cause of feast and famine in technology production down through the ages?

When you look at marvels like the Coliseum in Rome, on which construction started in 70 AD, and the many other large structures that survive from the Roman Empire, it is clear that Rome had plenty of engineers. Similarly, buildings like Notre Dame Cathedral in Paris (1163 AD), Westminster Abbey in London (1050 AD), and similar medieval structures around Europe suggest that engineers were again available 1,000 years later. One the other hand, the Dark Ages (476-800) saw few if any large construction projects in Western Europe. While this was a period when there was no Roman (or later Holy Roman) Empire in the west, weather may have played a larger role in the availability of educated engineers.

The Roman era was characterized by warmer drier weather, the Dark ages by colder wet weather, followed by the Medieval Warm period (900-1300 AD), and then the Little Ice Age (1300-1850 AD). When weather favors food production, fewer people are tied up in hand-to-mouth farming. That frees up some people and should result in more engineers, masons, and other specialists. Accordingly, one would expect more large construction projects and fewer wars when food is plentiful. Cause & Effect?

Seasons, Climate, and Ice Ages

Weather is what you see when you look out the window in the morning. Is it raining or sunny?

Seasons are periods of similar weather that extend over a period of months. Winter or summer?

Climate lasts hundreds of years and usually involves geography running east-west. However, interaction with nearby bodies of water can result in a climate that one would expect to find south of that latitude. A good example is the shoreline all around Lake Michigan. Gardeners know that the warmth of the lake moderates the winter enough to allow plants that could not survive 15 miles farther inland. Similarly, ice and snow are rarities in Ireland which is between the 51st and 53rd parallels that also run through Hudson's Bay in Canada and the Aleutian Islands.

Ice ages and warm periods between them are the extremes of climate change that occur in roughly 100,000 year cycles. Surprisingly, out of every 100,000 years, only about 20,000 are in warm periods. The other 80,000 are in the freezer!

All of the above are caused by the Earth's interaction with the sun, and moderated by the oblique angle of the axis of rotation of the globe, precession of the celestial pole, and the non-circular orbit around the sun. Together they are known as the Milankovitch Cycles (Fig. 5). They contribute to weather, seasonality, climate change, and ice ages, but do so over vastly different time frames.

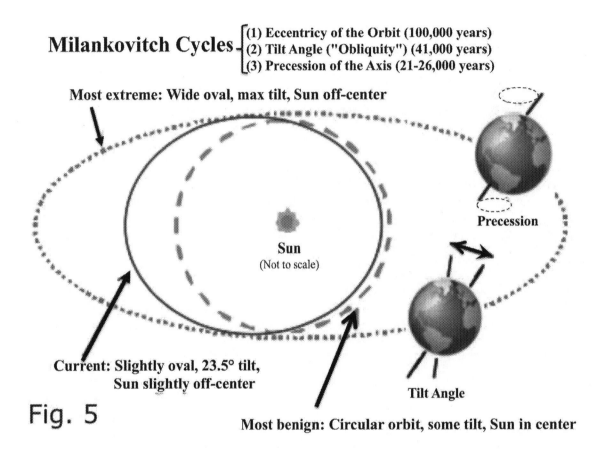

Milankovitch Cycles
(1) Eccentricy of the Orbit (100,000 years)
(2) Tilt Angle ("Obliquity") (41,000 years)
(3) Precession of the Axis (21-26,000 years)

Most extreme: Wide oval, max tilt, Sun off-center

Precession

Sun
(Not to scale)

Current: Slightly oval, 23.5° tilt,
Sun slightly off-center

Tilt Angle

Fig. 5

Most benign: Circular orbit, some tilt, Sun in center

Earth's tilted axis & seasons

The primary cause of seasonality (i.e. Spring, Summer, Fall, Winter) is the tilted or oblique angle of the Earth's axis of rotation relative to the plane of its orbit around the sun. Distance from the Sun in winter Vs. summer is a minimally significant factor. Earth is 93 million miles from the Sun, and the diameter of Earth is only 8,000 miles, so in a circular orbit, the maximum difference in distance from the Sun would be a mere 0.0086%. Do the math.

If the Earth's axis of rotation was perpendicular to its orbit, and the orbit was circular, there would be no seasons. It would always be hot at the equator and cold at the poles, where snow would accumulate into glaciers on land. Those glaciers would grow to be miles thick, because it would be a one-way trip for the water in the snow. Sea levels would drop around the world and deserts would be larger than at present. Not a good deal for anybody.

In spite of Earth's tilted axis, the north pole is pointed at Polaris in all seasons. Thus, when one hemisphere is pointed towards the Sun, the other hemisphere will be pointed away from it, i.e. summer Vs. winter, long Vs. short days.

In an earlier chapter we saw how the Sun shining to the bottom of an Egyptian well on the Summer Solstice helped Eratosthenes calculate the circumference of the Earth two centuries BC. Syene was on the Tropic of Cancer and the Sun was 90° at noon that day.

However, equal sunbeams impinging on the Tropics of Cancer and Capricorn on June 21 would be spread over vastly different areas, 40% more in the case of the latter (Fig. 6). It would be an oval shape rather than a circle due to the shallower angle of sunlight at that latitude. You can demonstrate this with a flashlight and a globe.

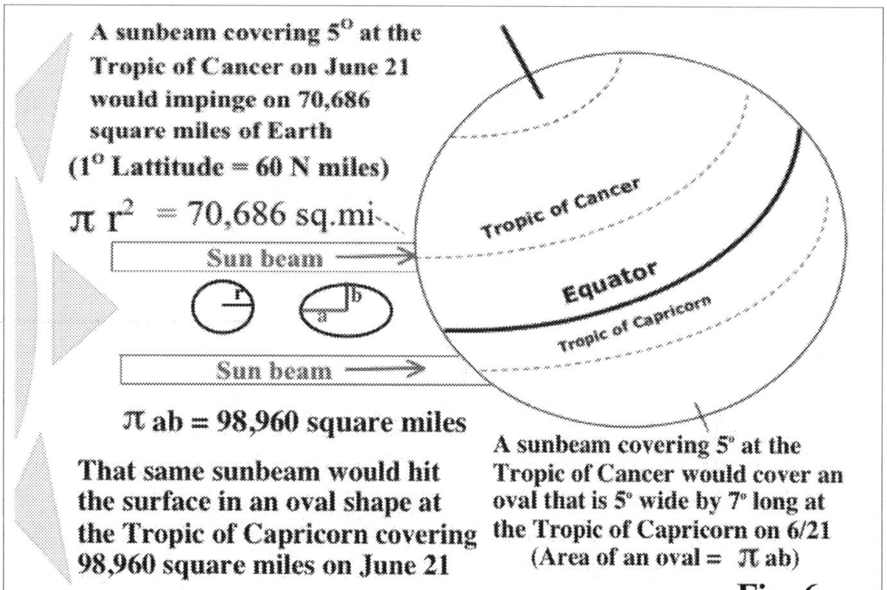

According to the Milankovitch Cycles, obliquity varies over a 41,000 year cycle. The greater the tilt angle of the Earth's axis of rotation, the greater will be seasonality and differences between the hemispheres.

The Earth is a massive gyroscope, and like any gyroscope, precession of the axis of rotation occurs. The net effect of precession of the Earth's axis is that it will no longer point at Polaris in a few thousand years time. I won't worry about it.

Precession of the Earth's axis of rotation occurs in 21-

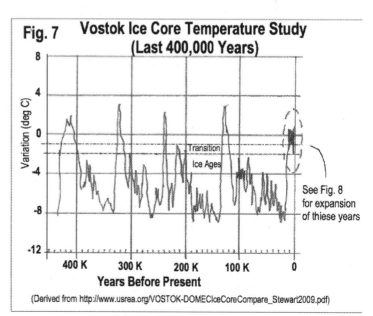

A sunbeam covering 5° at the Tropic of Cancer on June 21 would impinge on 70,686 square miles of Earth (1^O Lattitude = 60 N miles)

$$\pi\, r^2 = 70,686\ \text{sq.mi.}$$

Sun beam →

Sun beam →

$$\pi\, ab = 98,960\ \text{square miles}$$

That same sunbeam would hit the surface in an oval shape at the Tropic of Capricorn covering 98,960 square miles on June 21

A sunbeam covering 5° at the Tropic of Cancer would cover an oval that is 5° wide by 7° long at the Tropic of Capricorn on 6/21 (Area of an oval = $\pi\, ab$)

Fig. 6

26,000 year cycles. You might think that it would make no difference climatically, but the sun is not at the geometric center of the slightly elliptical orbit. In the Analemma discussion in the chapter *Navigation is Applied Geometry* we learned the Earth is closer to the sun and orbiting faster during our winter, so precession might contribute to differences between the hemispheres.

Eccentricity is the Primary Cause of Ice Ages

Eccentricity of the Earth's orbit around the Sun occurs in 100,000 year cycles and refers to the range of shapes that the orbit takes in the cycle. The most benign shape is a circular orbit and we have fairly close to that today, although the Analemma demonstrates that the Sun is not at the center of Earth's orbit around the Sun, and that orbit is somewhat elliptical.

The most extreme case is where the orbit is in the shape of a wide oval with the Earth closer to the sun at one end than the other. That opens the door to a new ice age. The other two effects, obliquity and precession, which occur in vastly different but overlapping time frames, add to the problem and results in the step-wise passage through an Ice Age seen in ice core studies.

Fig. 7 Vostok Ice Core Temperature Study (Last 400,000 Years)

Variation (deg C)

Transition
Ice Ages

See Fig. 8 for expansion of thiese years

400 K   300 K   200 K   100 K   0
Years Before Present

(Derived from http://www.usrea.org/VOSTOK-DOMECIceCoreCompare_Stewart2009.pdf)

As can be seen in Fig. 7, four ice ages have occurred in the last 400,000 years, and the current warm period began a little under 12,000 years ago. It had started 2,000 years earlier but suddenly reverted when the glaciers advanced again. See the blip in Fig. 8 circa 13,800 BC. That period was when the saber-toothed tiger, the wooly mammoth, America camel, dire wolves, giant sloth, woolly rhinoceros, short faced bear, and the American horse all became extinct. So something nasty happened.

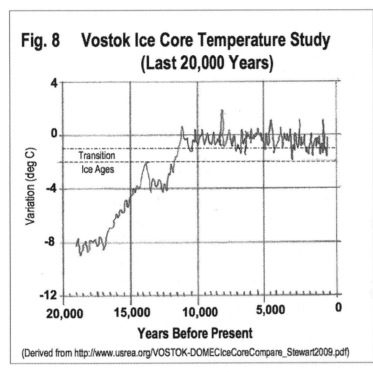

Fig. 8 **Vostok Ice Core Temperature Study (Last 20,000 Years)**

(Derived from http://www.usrea.org/VOSTOK-DOMECIceCoreCompare_Stewart2009.pdf)

One hypothesis is that glaciers covering all of Canada and the USA down to Chicago and New York receded north until the St. Lawrence River opened up. Until then, all water between the Rockies and Appalachians flowed down the Mississippi Valley to the Gulf of Mexico. When the St. Lawrence emerged from under glaciers a new path opened for melt water to flow into the North Atlantic. Being fresh, it had lower density than salt water and did not sink, thereby cutting off the Atlantic Conveyor.

Loss of the transfer of heat from south to north by the Atlantic Conveyor dropped temperatures and the glaciers advanced south again. That is a plausible explanation for why the glaciers stopped retreating and advanced again, but why did the animals and many types of plant life go extinct? They had already survived the ice age.

I prefer a second hypothesis, which has a meteor striking North America. It would have caused massive melting and flooding out through the St. Lawrence, thereby shutting down the Atlantic Conveyor just like the first hypothesis. As with the meteor which wiped out the dinosaurs 65 million years earlier, this meteor could have wiped out most life in North America at the time, and starved the survivors. But where is the impact crater?

Scientists are looking for evidence of this meteor strike. How much crater would you expect from a meteor that hit a two mile high glacier or exploded in the air? Were saber-toothed tigers parboiled by the emitted steam? How much water is there in a 2-mile thick glacier covering Canada and 1/3rd of the USA? Enough to raise sea levels by hundreds of feet? Let us do the math:

If a 2 mile thick glacier extended south to Chicago and New York and from coast to coast, then there would be 2 miles x 1,000 miles (north to south) x 3,000 miles (east to west) = 6,000,000 cubic miles of water tied up in the glacier. Divide that by the 129,000,000 square miles currently covered by oceans and you get 0.0465 miles as the depth of that water if it were spread out over today's oceans. (It is!) Multiply that by 5,280 feet/mile and you get 246 feet.

Thus, the water trapped in that 2-mile thick glacier would be enough to raise the sea level *in the current dimensions of the oceans* by 246 feet. However, if you were to take that much water *out of the ocean,* the continents would be much larger than they currently are and the surface of the Earth covered by oceans would be correspondingly less. Add the same amount of water to much smaller oceans and the depth increase would be *more* than 246 feet. This explains why a type of coral, which only grows in shallow water, can be found in 300 feet of water off the coast of Florida. At the end of the last Ice Age, that coral was in shallow water. Similarly, any human settlements that existed on the coast at that time would now be submerged under hundreds of feet of water. It's no wonder that we have not found them! The

84

Clovis Culture in New Mexico, which has been found by archaeologists, also appears to have disappeared suddenly about 13,800 years ago.

Strange Cause & Effect

How come Native Americans never invented the wheel? You probably remember the Sesame St. song: *"One of these things is not like the others; One of these things just doesn't belong."* Take a look at the four pictures.

Top left is a chariot Ramses used to chase Moses circa 1,200 years BC. Top right is Achilles at the Battle of Troy about the same time. Bottom left is Queen Boadicea of the Iceni who almost threw the Romans out of Britain in the first century AD. Famous for going into battle topless and with swords on the hubcaps of her chariot, she gave rise to the term 'bodacious lady', meaning 'outrageously arrogant & uninhibited'. Bottom right is an American Indian dragging a travois behind the horse circa 1870 AD.

If other societies had chariots thousands of years earlier, how come Native Americans never invented the wheel?

Prior to the arrival of the Spanish

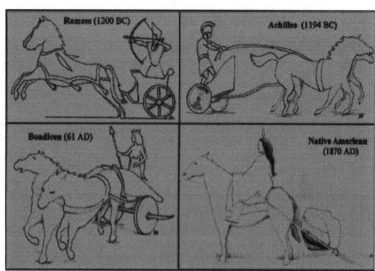

Fig. 9 One of these things is not like the others . . . .

conquistadores and the other Europeans that followed them, there were no horses, mules, or donkeys in the Americas. Although both horses and camels had existed on the continent, both were extinct for thousands of years before Columbus arrived. While the llama was used in South America, you can't ride a llama. Also, they never made it across the jungles of the Isthmus of Panama into North America. The closest thing to a 'beast of burden' in North America was the dog. Dogs were also the only animal domesticated by the Indians, and were used to drag a travois similar to the one in the picture. Unlike the water buffalo used in Southeast Asia, the North American bison could not be tamed.

With no beast of burden, there was no driving force to invent a cart, or a wheel, or a paved road for the cart, etc. Something caused those animals to go extinct and locked Native Americans in the Stone Age. My bet is on a meteor circa 10,000 BC. Now that is Cause & Effect that you were not expecting! (For more read *"Guns, Germs, and Steel"* by Jared Diamond)

Chapter 19   **Global Warming – Cause & Effect**

Consensus or Con-Job?

    The last chapter suggests that we are between Ice Ages, so we are either warming or cooling. But which is it? Before we get into that, I want you to answer a question: Which should you fear more, having Miami weather in Wisconsin or having North Pole weather in Wisconsin? I suspect that you will agree, Miami is preferable, but the question was "Which do you <u>fear</u>?" not "Which do you prefer?" The difference is significant.

    If you want to get people stirred up over some issue, impending catastrophe is more likely to get attention than something that most would consider beneficial. It may surprise you to learn that global warming was preceded by ice age alarmism. The June 24, 1974 issue of *Time Magazine* described how scientists had found evidence of global cooling that had been underway since 1940. *Newsweek* followed 9 months later with the graph in Fig. 1.

    The 1975 consensus among climatologists was that we were entering a new ice age. Desperate solutions included Swedish climatologist Bert Bolin's suggestion to burn more coal and oil in order to generate greenhouse gas that would stave off the anticipated ice age. Bert Bolin became the first chairman of the UN's IPCC (Inter-governmental Panel on Climate Change), which now hysterically warns of impending doom due to warming. Anthropogenic (i.e. man-made) global warming (AGW) alarmism is the claim that man-made CO_2 is causing Earth to heat up, putting life as we know it (and polar bears) at risk.

    Alarmists conveniently ignore the earlier ice age scare and the fact that Greenland must have been warmer during the Medieval Warm Period, when Vikings farmed there. Also, somehow polar bears survived.

    AGW alarmists not only ignore the Medieval Warm Period, they tried to write it out of history! While the first IPCC Report in 1991 included a graph of temperature over the last 1,000 years which depicted the Medieval Warm Period (900-1300 AD) and the Little Ice Age (1300-1850), the 2001 IPCC Report adopted the "Hockey Stick" graph with temperature in a flat line from 1000 to 1850 (Fig. 2). This was

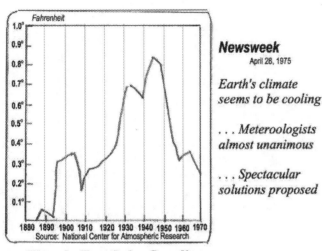

Newsweek
April 28, 1975

Earth's climate seems to be cooling

. . . Meteorologists almost unanimous

. . . Spectacular solutions proposed

Fig. 1 Earth is Cooling

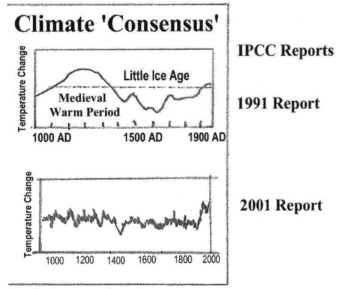

IPCC Reports

1991 Report

2001 Report

Fig.2 Hockey Stick Hokum   (WSJ 7/14/2006 p.A12)

subsequently shown to be a blatant fraud.

More evidence of IPCC having gone over to the dark side lies in how it treats dissenting results and in its rules, which ensure that IPCC reports serve political agenda rather than science. Scientists were told that their dissents were deleted from the final report "in order to ensure that it conformed to the policy makers' summary".

Should not a summary conform to a final report rather than *vice versa*? AGW skeptics frustrated by false claims of consensus among scientists were vindicated when the Climate gate emails revealed corrupted ethics and scientific methods.

If a 'reverse prism' (Fig. 3) could be used to dissect AGW alarmism, one would find that it serves many agenda. Motives include greed, environmentalism, ego, anti-capitalism, and political goals that span the spectrum from far left to far right. Environment is the one most poorly served by alarmists.

The Green Party is quite powerful in Europe. Many of its members are known as 'Watermelon Greens' because they are green on the outside only and red on the inside. Watermelons are more interested in redistribution of wealth, 'leveling the playing field', or 'one-world government', than in the environment.

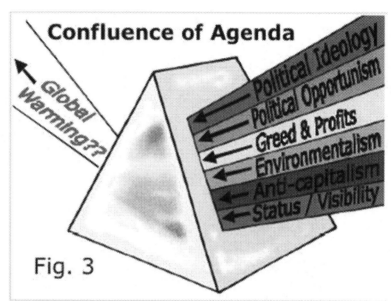

An example of their environmental impact is the focus on bio-fuel. It actually increases the amounts of oil used and greenhouse gases generated, as well as causing inflation in the cost of food. Pursuit of carbon-offset subsidies via bio-fuel crops is causing rain forest destruction, extreme weather, droughts, & desertification.

The IPCC's Kyoto Protocols set emission targets for countries, supposedly for environmental reasons, although politicians clearly think otherwise. French President Jacques Chirac said: "Kyoto represents the first component of an authentic global governance", EU Environment Minister Margot Wallstrom said: "Kyoto is about the economy, about leveling the playing field for big businesses worldwide". Sen. Kirstin Gillibrand (D-NY) said that trading in carbon permits could help New York by becoming the world's largest commodity market. UK Prime Minister Margaret Thatcher's Tory predecessor was kicked out of office by a miner's strike and the Arab Oil Embargo, so Thatcher set out to make coal & oil look bad and nuclear power look good. However, her agenda was hijacked by the 3 Mile Island accident and Chernobyl.

The Cap & Trade Bill (re-branded as the American Power Bill) is primarily a way to impose a stealth tax. It would have a negative impact on American businesses. What Watermelons did not achieve with legislation for Cap & Trade, they partially achieved by executive fiat when the EPA declared CO_2 "a dangerous pollutant". The Copenhagen Treaty called for One-World Government and wealth transfer from rich to poor countries.

Careers, crop subsidies, reputations & bureaucracies now depend on there being a climate crisis. Al Gore will continue to pocket his $300,000 presentation fees. He may also reap a windfall in carbon trading fees from a company he co-founded, Generation Investment Management (GIM). I wonder if he buys carbon permits to make his lifestyle 'carbon neutral'. GIM co-owns the Chicago Climate Exchange.

Correlation is not Causation

The most convincing part Al Gore's movie, *An Inconvenient Truth,* was where he showed a correlation of temperatures and CO_2 levels from the Vostok ice core study. Gore got a laugh when he asked: "Do they ever fit together?" The audience could see that they did (Fig. 4).

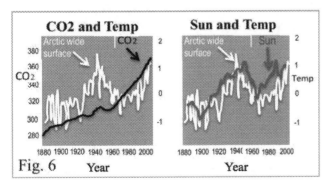

Fig. 4 Correlation between Temperature & CO_2

However, he was guilty of either 'presentation fraud' or ignorance, because he did not point out that the scale of the graph hid a large time gap between the two tracks. Presentation fraud occurs when a presenter deliberately provides incomplete data.

Gore should have pointed out that temperature peaks preceded CO_2 peaks by 800 years. When the data is graphed over a shorter period it is clear that temperature declined for hundreds of years while CO_2 levels were going up (Fig. 5). Oceans and volcanoes are the largest sources of CO_2, and less CO_2 can be dissolved in warm water than in cold water. The size of the oceans is why it takes so long to warm up and release CO_2, or cool and retain it. This multi-century gap between temperature peaks and CO_2 was replicated, and showed that CO_2 <u>followed</u> temperature long term, rather than *vice versa* as claimed by alarmists.

On the issue of presentation fraud, the atmosphere of Venus is 97% CO_2 and its average temperature is hot enough to melt lead, while that of Mars is 95% CO_2 and is so cold that water cannot exist as a liquid. So Earth's 0.038% CO_2 is probably not significant, right? If I left it at that, I too would be guilty of presentation fraud. It is significant that Venus is the 2nd planet from the Sun and has a dense atmosphere. Mars is the 4th planet from the Sun and has a very thin atmosphere. Deliberately leaving out data that conflicts with your argument is 'presentation fraud'.

Short term, temperature correlates well with solar activity and not with CO_2 levels (Fig. 6). Similarly, rising temperatures during the Great Depression and falling temperatures in WW2 and the rebuilding thereafter suggest that man-made CO_2 did not drive temperature in the short term.

Bottom line, both long term and short term, Earth's CO_2 levels have not driven global temperatures with one possible exception: Snowball Earth. 650 million years ago the Earth was covered with snow from pole to pole and solar energy was reflected back into Space. The sky would have been cloudless because all the water was trapped in the snow. However, volcanoes still erupted, spewing ash and CO_2 over the reflective surface. With no rain to scrub the CO_2 out of the atmosphere, the CO_2 level in the atmosphere was estimated to reach 11–12%. The greenhouse gas effect of CO_2 at that level probably helped capture

88

enough heat from the Sun for the green planet we now live on to emerge. Today the CO_2 level in the atmosphere is 0.038%.

To bring this chapter to a close, correlation is not causality and consensus is not science. I have provided some data and hypotheses in which I believe. But they are unproven. Keep your eyes and ears open for proof that may or may not emerge. That is the nature of science. A hypothesis is advanced and research is done to prove or disprove it.

Consensus is not science because it does not add to the body of knowledge. Ptolomy came up with a useful but wrong concept that the sun and stars all rotated around the Earth. Copernicus and Galileo risked their lives challenging it a 1,000 years later. Consensus wasn't science then and it is not science today, but faulty science does a lot of damage when it is accepted by a scientifically illiterate press and public and is not challenged by other scientists.

Bottom line, warmer is definitely preferable and is not to be feared.

Chapter 20   STEM Q&A – Weather & Climate

Questions	Answers
What causes deserts at 30° N and S of the Equator? (a) Hot air rises at 30° and descends at the Equator (b) Hot air rises at Equator and descends 30° N &S (c) Ancient locals killed off the trees at 30° N & S (d) Distance from the ocean is the key factor	**(b) Hot air rises at Equator and descends 30° N and S** Hot air rises at the Equator and loses moisture as it cools. This is why you have rain forests at the Equator. That dry air comes back down circa 30° N and S, which is why deserts occur at those latitudes.
In what direction are winds around a low pressure area? (a) Clockwise north of the Equator (b) Counter-clockwise north of the Equator (c) Counter-clockwise north & south of the Equator (d) Depends on proximity to major bodies of water	**(b) Counter-clockwise north of the Equator** The opposite is true south of the Equator. Around a high pressure north of the Equator, the winds are clockwise.
What causes seasonal changes in Monsoon winds? (a) Uneven heating of land & sea in winter Vs. summer (b) Earth's rotation combined with winter Vs. summer (c) Less CO_2 is produced in the winter (d) More CO2 is produced in the summer	**(a) Uneven heating of land vs. sea in winter vs. summer** India warms in summer and rising air is replaced by cold air flowing on-shore from the sea. In winter warmer water causes the opposite effect and winds flow in the opposite direction.
When do airplanes fly better, in warm air or cold air? (a) Warm air (b) Cold air (c) Temperature makes no difference	**(b) Cold air** Cold air is denser per Charles' Law, so an airplane pushes more air molecules down at any speed. This means shorter take-offs and better performance.
Does the composition of air change as you go up? (a) Yes - at high altitudes ratio of O2 to N2 is lower (b) No - composition unchanged, but pressure is lower (c) Yes and no, water content changes, but the ratio of other gases to each other remains the same	**(c) Yes and no** While clouds demonstrate differences in H_2O content, the ratio of O_2 to N_2 is unchanged. However, reduced pressure at altitude requires supplemental oxygen for pilots.

Questions	Answers
What causes the periodic Ice Ages? (a) Variation in Earth's orbit and axis of rotation (b) Meteors and asteroids hitting the Earth (c) Don't know, don't care. It won't affect me. (d) The Sun cools off periodically	**(a) Variation in Earth's orbit and axis of rotation** Per the Milankovitch Cycles, eccentricity in Earth's orbit varies in a 100,000 year cycle, the oblique angle of Earth's axis of rotation varies in a 41,000 year cycle, and the axis precesses over a 21-26,000 year cycle.
Why isn't the air sucked off by the vacuum of Space? (a) It is, but it is being replaced constantly (b) Earth's gravity has a strong grip (c) Earth's grip on the atmosphere is very loose	**(b) Earth's gravity has a strong grip** The Equator rotates to the east at >1,000 mph, so a wind that fast might be expected, since the atmosphere is not as tightly bound as mountains and continents. But the Trade Winds are much slower. So the atmosphere is tightly held, but with some slippage. That slippage is the Trade Winds.
What causes the four seasons? (a) Earth is closer to the Sun in the Summer (b) Earth's tilted axis of rotation relative to its orbit (c) Earth's elliptical orbit around the Sun (d) The Sun is not at the center of Earth's orbit	**(c) Earth's tilted axis of rotation relative to its orbit** In the northern Summer the Sun's beams are more directly oriented to the northern hemisphere and hit the southern one at a lower angle
What causes the Trade Winds? (a) Rising air over land sucks wind from seaward (b) Earth's rotation and replacement of rising air (c) Equatorial ocean currents flow in that direction (d) They are a function of winter Vs. summer	**(b) Earth's rotation and replacement of rising air** Warm air rises at the Equator and has to be replaced from north and south. At the Equator Earth is rotating east at >1,000 mph. Together they cause northeast and southeast winds north and south of the Equator respectively.
Preflight inspection includes checking fuel for water. How does H$_2$O get into the tank of a plane in a hangar? (a) It comes out of solution from the Avgas (b) Condensation from moist air (c) It is picked up in flight through leaky gas caps (d) It is produced by static electricity in the tank	**(b) Condensation from moist air** Warm air holds moisture. When it cools condensation can occur in a fuel tank. Heavier than gasoline and immiscible with it, H$_2$O accumulates at the bottom of the tank. This is why full tanks will accumulate less water.

Chapter 21

Mega-Historical Cause & Effect

Mega-history is history in 500 year chunks. In their eye-opening book, *The Great Reckoning,* James Dale Davidson and William Rees Mogg suggested that every 500 years or so the introduction of a new weapon so disrupts the status quo that the whole society changes, how it is governed, functions, how people live, laws, taxes, trade rules, etc. It has happened again in the last 20 years. The transition from the Industrial Age to the Digital Age is one result. Some countries were more ready than others. Ireland's Celtic Tiger was an example of a country that was ready, and the now defunct USSR is an example of one that could not survive the change.

Weapons Create 500 Year Epochs

 Stirrups lead to Feudal Era (knights/castles, then City States)

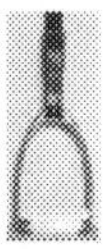

 Cannon gun ended Feudal Era, created need for Nation State

 Microchips obsolete Nation State, and foster the *Invisible Continent*

Fig. 1

The three transforming weapons in the last 1,500 years were the stirrup, the cannon gun, and the microchip (Fig. 1). Weapons? Yes, all were weapons and all transformed society. The stirrup enabled the Feudal Age. Cannon guns ended that and created the need for the Nation State. The microchip recently made the Nation State obsolete and fostered the Invisible Continent described in Kenischi Ohmae's book of the same name.

In the 9th century, when Europe was mired in the dark ages following the break-up of the Roman Empire, the introduction of the stirrup into Europe enabled armored knights to dominate anybody who could not afford a horse and armor. That would have included about 99% of the population. A knight in full armor attacking people on foot in a field would be like a tank attacking infantry. Without stirrups, full armor was not possible, so their introduction was key to the whole Feudal Era. For more on this topic read *The Great Reckoning* by James Dale Davidson and William Rees Mogg.

Enter the Age of Feudalism

Feudalism came into being in Europe anywhere horses could be used in battle. Disproportionate power concentrated wealth into a few families who built castles to protect their holdings. The Lord of the Castle effectively owned both the land and the serfs that lived around the castle.

"My lord's lord is not my lord" is a saying that described the relationship between serfs in a feudal fief, the lord of the castle, and a higher aristocrat or king. Because the serfs owed allegiance only to the lord of their castle, each castle was effectively a separate country. Their local lord might owe allegiance to a king, but that was <u>his</u> problem. It was like being a citizen of Racine but not of Wisconsin or the USA.

Today's gated communities with their governing councils and private police forces hearken back to feudal days. During Feudalism, cities surrounded themselves with walls and some became independent countries. The most successful ones were those - like Venice & Amsterdam - who were involved in international trade and had access to the sea.

Back in the Feudal Era the only effective way to take a castle was to starve out its inhabitants, and they tended to be stubborn because the penalty for losing was serious. When Carrickfergus Castle fell to a siege in 1317, it was not until its occupants had eaten all eight of their Scottish prisoners. The rules were different back then. Such differences are what I am trying to get across, new rules for a new era, the Digital Age and the Invisible Continent.

Blarney Castle is a ruined castle like many others all over Ireland. When I was growing up we used to play Cowboys and Indians in one near my home. However, Blarney Castle is famous for the Blarney Stone. Every year thousands of Americans pay for the privilege of kissing the Blarney Stone. It is an odd colored stone inside the parapet of Blarney Castle. The bars that this tourist (Fig. 2) is hanging onto and those across the bottom of the hole were added for the tourists. They certainly were not there in the old days when prisoners were dangled in the same position and invited to tell all that they knew. With the alternative being a 13 story drop to the rocks below, many of them got the gift of the gab. There was no magic involved! It was pure Cause & Effect. That being said, the Gift of the Gab

Kissing the Blarney Stone give you the "Gift of the Gab"

Fig. 2

bit differentiates Blarney from other ruined castles and adds millions of dollars to the local economy every year.

Feudal Age castles did not have bathrooms as we know them today. I have no doubt that the soldiers used to go to the bathroom down through this hole, which may be why the Blarney Stone was discolored in the first place. But if it was called the *Pee Stone* I don't think tourists would pay to kiss it. Enough nostalgia, back to the 2nd and 3rd Epoch Changing Weapons, the cannon and microchip.

Castles used to be secure

Blarney Castle could hold out when attacked with ladders and catapults, but it was no match for cannon guns. An early use cannon guns was during the invasion of Italy by Charles VIII of France in 1494. A fortress that had withstood a 7-year siege just a few years earlier had its walls knocked down in 8 hours. Feudal castles had high walls, but they were not very thick or sloped. Cannons could breach them easily.

The first use of canon guns in warfare was in the capture of the Christian city of Constantinople by the Muslim Ottoman Empire in 1453, giving Muslims a strong foothold in Eastern Europe just before they were driven out of what is now Spain and Portugal in Western Europe.

At the end of the Feudal Era there were 500 sovereign "states" in Europe. There were only about 25 left in 1914 at the start of WW l. Small kingdoms and city states needed to combine for mutual support against powerful enemies with artillery. Nation states became the norm, and the power equation changed from defensive to offensive so effectively that countries which led the Gunpowder Revolution controlled most of the globe.

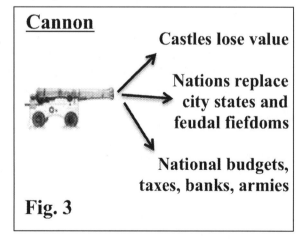

Cannon

Castles lose value

Nations replace city states and feudal fiefdoms

National budgets, taxes, banks, armies

Fig. 3

The switch from the Feudal Era to Nation-States created parliaments and central governments. A lot of power shifted to commoners and the middle class grew in size and importance. Society changed just in time for the Industrial Age. This began when steam engines automated factories in the mid 1800's.

The arrival of cannons started an arms race which would bankrupt the city states. The arrival of the next weapon, the microchip, would bankrupt an empire, specifically, the USSR. Will it also bankrupt the American Empire?

500 years have passed and it happened again. Microchips and what they enable are the modern equivalent of Charles VIII's cannons. A Stealth Fighter used one smart bomb to destroy Iraq's Defense Ministry. WW2 technology would have needed 9,000 bombs. A Taliban fighter with a shoulder mounted (US supplied) guided missile could take out the highest tech Soviet helicopter.

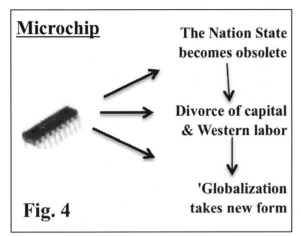

Microchip

The Nation State becomes obsolete

↓

Divorce of capital & Western labor

↓

'Globalization takes new form

Fig. 4

Society will change as surely as it did when the stirrup and the cannon gun were introduced.

The fall of the Berlin Wall in '89, and of the Soviet Union soon after, marked the end of the need for the Nation State. THAT triggered the divorce of capital and western labor with far reaching consequences that continue to emerge.

In this new era urban terrorism is a greater threat than nuclear war. 'Water-boarding' became a campaign issue in US elections. The fact that it was even being discussed as a political issue was a measure of the change since WW2. It was a whole new ball game with new rules.

With less need for strong nations, there is more need for strong municipalities with cameras on streets. Don't be surprised to see some countries split into their pre-cannon gun regions. Like Yugoslavia and the USSR have already.

Since capital can now chase skilled labor and lower costs anywhere in the world, the challenge of having a corporation that is accountable to community is magnified.

Being a citizen of the USA can be a liability.

I traveled all over the world for work, and used to have a US passport in my left pocket and an Irish passport in my right pocket, in case somebody stood up in the plane and yelled *"This is a hijack!"* I often had a Maple Leaf flag in my lapel, as I could easily pass as either a Canadian or Irish when encountering Anti-American crowds. One night in Istanbul I was asked *"Are you American?"* 3 times. It was the week in 1986 when America shot down two Lybian fighters in the Gulf of Sidra and was a mite scary. I was Irish that night.

Don't be shocked. The globalized economy mocks the assumption of shared political values that supposedly unite people in the nation state. Politicians talk about the "Two Americas", or the "1% Vs. the 99%". Both slogans only further divide us, and offer few solutions to the new reality. Too many low skill jobs will end up being done by technology, so the solution to the problem is to provide workers with skill sets that are valued today and will still be valued ten years from now. That is why STEM education is so essential.

A vivid example that comes to mind is garbage pick-up on my street. It used to require two trucks, one for regular garbage and the other for recycling, and both trucks had a driver and a man to handle the bins. Today there is one truck and one worker, the driver. He never leaves the truck. He zooms down the street and stops next to my bins. An articulated arm grabs the specially designed bin, dumps its contents into the truck, and replaces the bin. In less than ten seconds the truck is on its way to my neighbor's house. The whole operation is so fast that the same truck comes back later in the day to pick the recycling garbage. Bottom line, the three jobs that were eliminated had not been sent to China, they were rendered obsolete by technology.

All that weapon stuff is a backdrop for the stage, but what is the play and how does this relate to the Digital Age? The reason for presenting this mega history is to explain why great change is inevitable. In my opinion, Congress is getting it wrong by focusing on off-shoring of jobs. That is Industrial Age thinking. They should look for ways to attract capital, not scare it off. America needs capital more than capital needs America. That is the new reality.

High tech is the main cause of income inequality

High technology eliminates middle and low end jobs. How many bank tellers were replaced by ATM machines? Countries and workers that adapt to the new reality will be the winners.

In Alabama Toyota has to use 'pictorials' - comic books - to teach semi-literate workers how to use high-tech equipment. Poor education was one reason why Toyota passed up huge financial incentives and located a new SUV plant in Canada.

High Tech + Mobile $'s ➔ New Rules
* Automation cuts more jobs than off-shoring
* Technology content of jobs has risen
* Invisible Continent of Judas Economy?
* The easier it is to fire you, the more likely you are to get a job in the first place
Fig 5

Millions of cheap laborers resulted from the demise of Communism in Eastern Europe and China's shift to a market based economy. This resulted in "deflation" in high labor content manufacturing. Governments and societies that don't acknowledge the twin realities of mobile capital and high technology face higher unemployment.

People may live in Wisconsin, but they work in what Kenichi Ohmae called *the Invisible Continent*. Wolman and Colamosca called it *the Judas Economy* in their book. They are two faces of the same coin.

The European Social Model guarantees chronic unemployment because it is too risky to hire people. Young people in France rioted over a proposed change in the rules relating to job security. The government wanted to make it easier for a corporation to shed unneeded labor. In effect, they were rioting over losing the right to keep a job that they were never going to get. The Digital Age demands more flexibility than we are used to. What the rioters were missing is a simple fact of Cause & Effect:

The easier it is to fire you, the more likely you are to get a job in the first place.

When governments make it so expensive to fire an unsatisfactory employee, then corporations are reluctant to take the risk of hiring. That helps nobody.

Today well-educated young Germans accept unpaid internships, hoping for a rare shot at a permanent position. Most Europeans still don't get it. Ireland and the UK don't use this social model, as there is nothing social about chronic unemployment. Why use that as a model?

4 Economies of Kenichi Ohmae's Invisible Continent (Fig. 6)

I grew up in the Visible Economy. An example is walking into Barnes & Noble to buy a book. Many old US corporations in the Visible Economy are now in better shape than their European counterparts because they restructured aggressively in the 1980's and again during the dot.com bubble.

If you ever bought books from Amazon.com you experienced the **Cyber Economy**, where >50,000 people now make their living on E-Bay. Less than 30 years old, the

4 Economies of *Invisible Continent*	
Visible:	Buy in Barnes & Noble
Cyber:	Buy from Amazon.com
Borderless:	Where is Amazon?
High Multiples:	Much higher profits
	Fig 6

Cyber Economy made it possible to find and buy things that in earlier times would have been thrown into a landfill. That is deflationary. The US is in good shape in the Cyber Economy because all the key platforms are American, as is the key intellectual content.

The 3rd dimension is the Borderless Economy. Where is Amazon located? **(a)** I don't know. **(b)** I don't care. **(c)** I'll bet that it is not in a high tax neighborhood.

American multinationals operate globally and US consumers are used to buying the cheapest and the best products from anywhere in the world. The US is in good shape there.

A Celtic Tiger example of the Borderless Economy is the $300M in taxes Microsoft paid to Ireland in 2004. If $300M was the 12.5% Irish corporate tax, then Microsoft sales in Ireland were $2.4B. That is either $600 for each man, woman, and child in Ireland - unlikely you'll agree - or MSFT is funneling sales in the rest of Europe through Ireland to avail of the low tax rate.

How much of that $2.4B counted as an export from the USA in the trade balance tables? I suspect very little, but the $2.4B is still owned by share holders of an American corporation.

48% of the profits of corporations on the S&P Index are from foreign sales. Furthermore, nearly 50% of our imports are goods that travel through intra-firm channels. As an example, when Johnson Wax imported a jar from a Johnson factory in China and then filled it with Glade in Waxdale, where did the profit go? To China or to Racine? The answer is 'Yes', but most of it went to Racine, WI.

I did my own taxes with Turbo Tax. I had a problem when Turbo Tax doubled the dividend for a municipal bond. It was not a problem at the federal level, but it was costing me in the state return. So I e-mailed their service center at 1:30 AM and Jonathan responded at 10:00 AM. His answer did not satisfy me, but he provided an 800 number.

I called and talked to Romeo. He did not sound American, but what does an American sound like? His English was good. Romeo spent 1/2 hour plugging the interest into box 8 and 9 on Form 1099-INT, following it though Schedule B, and finally into line 8b on Form 1040.

That is a lot of detail I know, but that is my point. He simulated the data that I entered and I was able to follow him on my own computer. He twigged the problem when I mentioned - in a purely throw away comment - the color of the numbers on Schedule B. I had both downloaded data from my broker and manually entered it when responding to the AMT test. Where was Romeo located? Don't know. Don't care. But he delivered. That's the Borderless Economy.

The 4th dimension is the High Multiples Economy. Basically, it explains how AOL, a start-up with the flakey finances of the dot.com era, could buy bricks & mortar stalwarts like Time Magazine and Warner Bros.

Example of *High Multiples* Economy

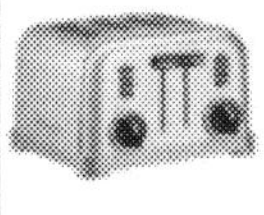

Toaster

List price:	$30
Incremental cost:	$5
Multiple:	6

Windows XP

	Disc	Download
List price (2006):	$200	$200
Incremental cost	50¢	5¢
Multiple:	400	4,000

Fig 7

While the other three are obvious, Kenichi Ohmae's "High Multiples" Dimension is the most difficult one to understand and is probably the most significant difference between the Industrial Age and Digital Age.

Think of an Industrial Age product like a toaster that has a list price of $30. The incremental cost to produce it in China is about $5. That covers the cost of bending the tin, coiling the heater wire, assembling and packaging it. $30 sale price Vs. $5 cost is a multiple of 6.

Compare that with a Digital Age product like Microsoft's Windows XP which was launched in 2001 and replaced by Windows Vista in January 2007. If you had bought Windows XP in December 2006 the list price was $200. Being 5 years old, the product's development costs had been paid off for several years, so its incremental cost would have been at most $0.50 if you bought the disk in a box. On the other hand, if you bought it by downloading it from the internet, the incremental cost would have been about 5 cents. That is a multiple of either 400 or 4,000. Compare that with a multiple of 6 for an old world product like a toaster.

This very high profitability of successful Digital Age products is why you have so many dot.com billionaires. Inevitably, it expands the gap between educated and uneducated people. While technology is increasing employer's appetites for some skilled jobs, it is diminishing it for assembly line type workers. Unfortunately, our public education does not reflect the demands of the Digital Age.

In addition to higher profits for those with a good handle on the digital purse, the Digital Age is deflationary, i.e. it drives down prices, which hurts the profit margins of those companies that are now up to the challenge. "Show rooming" has become a problem for Big Box stores like Best Buy, who find that customers come to their store to check out merchandise, and then buy it for a lower price on line. In cases like music CDs or movie DVD, the physical product may be eliminated entirely when downloading delivers the same benefit to the customer.

A common metaphor for society is a pyramid, with the wealthy plutocrat at the top and the grubby masses at the bottom. It was a useful image if you were trying to sell communism or socialism, but it is not very accurate, as some of the grubby masses are much worse off than others. Ammon's Turnip more accurately reflects society.

Ammon's Turnip

At the point on top you have the cognitive elite of the Digital Age, Bill Gates and other dot.com billionaires, the intellectual cognoscenti. Below that you have people like the Johnson or Kennedy families, then the wealthy professionals. The fat part is most of society. The point on the bottom is the uneducated poor. They are probably screwed no matter what. The best they can hope for is the minimum wage.

America's problem is represented by the 'Squeeze Zone'. The dashed line was the skill level for minimal success in the Industrial Age. When I started work in 1970 with my PhD, a neighbor's son, Joe, started work for the same company as a technician. Joe had just graduated from high school. He subsequently got drafted, served time in Viet Nam, and came back to the company where he became a forklift truck driver. 37 years ago Joe was above the dashed line. He is lucky he is not starting today!

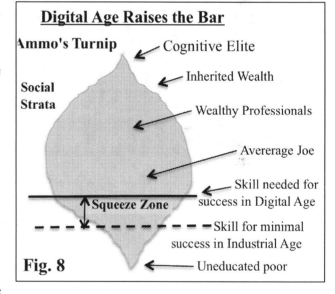

The solid line represents the skill needed for minimum success in the Digital Age. Everybody between the two lines will get squeezed by technological unemployment, and that will be a bigger problem for the US than it will be for less developed countries like pre-Tiger Ireland. If Joe was to start his career today he would be in the Squeeze Zone. Robotic pallet trains already do much of Joe's original job. Automation is a big job eliminator.

While the market for their managers will still justify bonuses, people in the Squeeze Zone will be measured against both automation and low cost, non-union labor here and overseas. UAW members are an example of the kind of people who will be most affected.

Digital Age Skill Sets

The education level of our workforce met requirements when most people were hired from the neck down in the Industrial Age. That is no longer true and unfortunately much of our public education system looks like the auto industry which was stuck in an outdated production model driven by the needs of employees rather than consumers. The consumers in schools are of course the students. Which do we want our schools to be more like, Detroit or Silicone Valley?

Today's students will compete for jobs in an environment where people are hired from the neck up. Failure means 'no job' or low wages that go with the surviving 'neck-down' jobs.

Did you know that 30% of high-school math students in the USA and 60% of physical-science pupils have teachers who didn't major in the subject or weren't certified to teach it. Therefore it should be no surprise to hear that 44% of science doctoral candidates in the US are foreign born, most of them on temporary visas. US students under-perform their peers in other nations in math and science achievement tests, and that raises worries about our future competitiveness. Worse still, 14-15% of young men age 20 and up in the USA lack a high school diploma.

Economic Growth Vs. Cause & Effect

Access to capital is a key driver of economic growth. That requires low taxes and property rights, plus the ability to use technology. Since Digital Age employers can locate anywhere, tax policies should welcome employers. Ireland's corporate tax rate of 12.5% is the lowest in the world and its workers are well educated. In 2004 Ireland with a population of 4 million got more investment ($7 Billion) from US corporations than China's population of 1 billion ($4 Billion), in spite of China's lower wage levels.

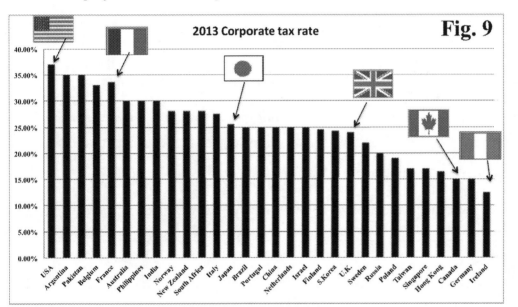

The result was the Celtic Tiger and the fastest growing economy in Europe. Why didn't that investment stay in the USA? Simple, even without counting state taxes, US federal corporate taxes are the highest in the world. If you had $100 million to invest in a new factory, would you invest it in the left side or the right side of this (Fig. 9) corporate tax graph?

Education – Cause & Effect

An educated work force is essential in the Digital Age. Workers compete on a global stage. The 'No Child Left Behind' law attempted to improve school quality in the USA via standardized measures, but did not define the level of proficiency required. That was left to states, possibly to satisfy Republicans' attachment to state's rights. According to a WSJ article, Wisconsin set the 8th grade reading passing level at the 14th percentile, whereas NC set its at the 71st percentile. A child moving from WI to high school in NC might have a problem.

Democrats did even more to castrate NCLB. The teachers union opposed all serious forms of school and teacher accountability. All corporate people know the axiom *"What gets measured gets done."* Our students need to be measured Vs. students in Europe/Asia. The threat of NCLB was that children at failing schools could get vouchers they would use to attend private schools. If schools had not been failing, no vouchers would be available or needed.

Taxes – Cause & Effect

Every penny that the government spends ultimately comes from a tax payer. Communism and Capitalism are fraternal twins that differ in <u>when</u> the government confiscates your money. Communism assumed that everything was owned by the government on behalf of the people, and resources were invested by a central authority. This was never successful, because there was little incentive for workers to do more and the central authority was never as knowledgeable as people closer to the issue at hand. Profit sharing was rare because there rarely were profits.

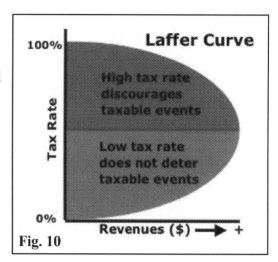

Fig. 10

In the case of Capitalism, the assumption is that individuals own the investment capital, whether it be a small farm or a steel mill, and those individuals are entitled to the resulting profits until tax time. That means there is more to redistribute. However, if the government's cut is then too high, incentive is reduced as was the case with Communism. If taxes are low, then the investor is encouraged to continue making profits. The Laffer Curve is a good explanation of this effect (Fig. 10). The Bush tax cuts in 2003 and Microsoft's first dividend are examples of it.

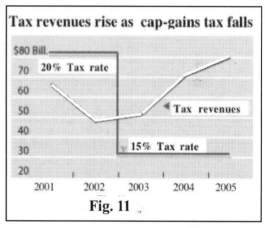

Fig. 11

While the left moaned about the Bush tax cuts in 2003 being "giveaways to the rich", the reduced taxes on capital gains from 20% to 15% resulted (Fig. 11) in a doubling of capital gains (from $269B in '02 to $539B in '05), and taxes revenue went from $49B in '02 at the 20% rate to $80B in '05 at the 15% rate. A 25% tax cut resulted in 63% more tax revenue!

The tax on dividends was reduced from effectively 38.6% to 15% and Microsoft paid out its first dividend, $32B in a single payment. 15% of $32B is $4.8B in tax revenue. 38.6% of $0 is $0. Why did Microsoft not pay dividends before the tax cut? Dividend money – which is from corporate profits – is what is left after already paying corporate taxes. Because shareholders then pay tax on dividends at their own marginal tax rate, dividends would be taxed twice, to a total of about 60%. They were still taxed twice, but now at 15% at the individual's level and not 38.6%. Government got more revenue, and so did the investor. This Win-Win was pure Cause & Effect.

Gap between Rich & Poor

Not only will the gap between rich & poor increase in the Digital Age, society may change as much as it did going into and out of Feudalism. Many jobs have been eliminated by technology and they are not coming back. While farm laborers or garment workers in a factory might have legitimate claims that their contributions are a major part of producing a product, this will be true for far fewer people in a digital environment where the product is computer code for the next version of Windows®. The notion that the factory of the future will have just two employees, a man and a dog is only partly a joke. The dog's role is to make sure that the man does not touch any of the machines, and the man's role is to feed the dog.

There are many examples where people have been eliminated almost completely from particular jobs. Examples include ATM machines which interact with customers 24/7 but have minimal contact with its own support staff. Most of the toll gates on a highway run automatically. When is the last time you saw a Lift Operator in a hotel?

What can society do about the technologically displaced workers? Will Ayn Rand's vision in *Atlas Shrugged* come to be true? What happened in the past that suggests a solution?

Chapter 22   STEM Vs. Paradigms

Paradigms are rules or patterns that control how people view something. Essentially, they are a rut in which our minds get trapped. They are not necessarily wrong, as was the case with Ptolemy, but they are always limiting. Ptolemy believed that the Sun, planets, and stars all rotated about the Earth (Fig. 1). His tables showing where planets would be at particular times were so convincing that his geocentric model went relatively unchallenged for over 1,000 years. Galileo was arrested for doing so with lots of evidence. One wonders how many similar paradigms control our thinking today and are susceptible to emerging technologies. One thing we can be sure of is that the proponents of 'consensus science' and 'conventional wisdom' will fight them tooth and nail.

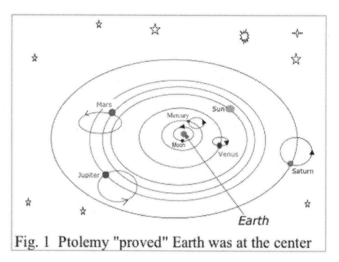

Fig. 1 Ptolemy "proved" Earth was at the center

The arbiters of paradigms

Who has/had the power to maintain paradigms? In the case of Galileo it was the Church. In many cases today it is an agency of the Federal government such as the EPA or FAA. They can be just as resistant to contrary evidence as the Church was in Galileo's day. The way corporations define their business is also a paradigm. For example, railroads were dominant when airplanes and buses were in their infancy. If railroad executives had seen themselves as being in the transport business rather than the railroad business, their corporations might not be in as much trouble today.

When managing groups of development chemists I would often ask them to investigate all aspects of a product and challenge every assumption, even aspects that were outside their supposed sphere of work. How were costs determined? How was the product to be tested to satisfy government requirements? Did those requirements reflect the latest science, or were they left over from earlier times? What was the impact of warehousing and distribution on package requirements? Had consumer habits changed in a way that impacted what they wanted from the product? Etc. Great opportunities could be uncovered by stepping "out of the box", but you first had to know the shape of the box.

One example that comes to mind related to how my corporation treated the cost of water. The corporate accounting system attributed a cost of \$0.000 / lb. to water. While water was indeed cheap, it turned out to be our most expensive raw material. Paradigm paralysis had blinded us to the way water drove up the cost of the total product.

The problem was partially due to compartmentalization. Chemists developed a disinfectant cleaner, package engineers and designers developed a package for it, distribution and accountants and many others would do their parts, and everybody had their own sets of paradigms. One problem was that the package cost roughly twice what the liquid inside cost. If we added 16 oz. of water to 16 oz. of disinfectant, the 32 oz. that resulted would have the same raw material cost but much higher packaging cost. However, if the consumer had diluted the original product 20:1 when using it, he/she could now dilute it only 10:1 to achieve the same level of disinfection, so the net cost/job would be higher and not lower. This suggested an opportunity.

Instead of adding more water, we took water out and concentrated the product by a factor of five. The consumer would dilute the concentrated product 100:1 rather than 20:1, and the package costs dropped to a fraction of what they had been. In addition, the package needed 1/5th as much warehouse space, 1/5th as many trucks to get it to market, so we were golden. Right?

Wrong! The Accounting Department's formula for calculating shipping and warehousing costs were a function of the <u>value</u> of the bottle of product, not the actual number of trucks and space in the warehouse. Being more concentrated, the value per ounce was significantly higher. The Accounting Department's formula had to change before we could launch a product that was much less expensive and far more environmental.

Now you might be wondering if a corporate pricing policy belongs in a book about STEM. However, it amounted to an extra arrow in the quiver of the development chemist. The M in STEM stands for math, and in this case the math formula involved in determining the cost of the technology drove its design. STEM is a continuum, and all aspects need attention.

Will Government Ever Admit Its Errors RE Global Warming?

Bio-fuel is made from corn based alcohol and was intended to replace some of the imported Arab oil. Energy independence and less "new" CO_2, what's not to like?

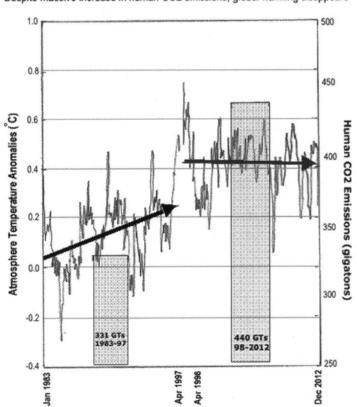

Fig. 2 IPCC/HadCRUT Dataset Confirms Global Cooling
Despite massive increase in human CO2 emissions, global warming disappears

Unfortunately, the math shows that we actually burn more oil to make, distribute, and use gasohol than if we used only Arab oil. So more CO_2 is produced. Additional unintended consequences are higher costs for animal feed stocks, thereby increasing the costs of tamales, chicken, and pork. Also, production of gasohol has a negative impact on ground water.

The globe is not warming (Fig. 2), the Arctic ice pack is not receding, and both Canada and Australia have acknowledged that they made a mistake when they jumped on the Global Warming bandwagon. But the EPA is still pushing this flawed agenda.

Will Dirigibles Come Back?

Dirigibles may or may not make a comeback. Key advantages are (a) fuel efficiency and (b) no need for an airport to handle large cargos. As long as you do not have to vent the gas, the lift provided by it only has to be paid for once and not every time you take off. Remote mines and areas ravaged by natural disasters could be served. Helium, unlike hydrogen, is nonflammable, and it is readily available today. A dirigible's key disadvantages include the huge size, the difficulty of handling them on the ground, and the paradigm effect of the *Hindenburg* disaster. Bottom line, the use of helium rather than hydrogen would have avoided that disaster, and concepts such as the Aeroscraft address the other disadvantages.

Nevertheless, paradigm paralysis is powerful and courageous investors will be needed. Dirigibles were almost erased from the scene. You probably know that the first trans-Atlantic flight was made by Alcock and Brown in a Vickers Vimy bomber. They left St. Johns, Newfoundland on June 14, 1919 and crash-landed in a bog near Galway, Ireland just under 16 hours later. While that clearly was a remarkable flight for its time, were you aware that a British Zeppelin, the R-34 flew from Scotland to New York just two weeks later (July 2-6, 1919)? Furthermore, R-34 departed New York on July 10 for the return flight, arriving in the UK after a flight of 75 hours and 3 minutes. Which was more impressive? The Vickers Vimy that flew between islands off the coasts of the two continents and crashed on landing, or R-34 which made a round trip to and from New York?

Design Limits

Sometimes design of a technology is constrained by arbitrary rules that are thousands of years old. The solid rockets boosters (SRBs) of the Space Shuttle come to mind. NASA engineers wanted them to be larger but were unable to have it so. The SRBs were manufactured in the Morton Thiokol plant in Utah and were transported to Cape Kennedy by rail. Those rail lines went through tunnels, and the size of the tunnels and what could be carried through them was dictated by the gauge of the rails, which is a strange number, 4 feet 8.5 inches.

Why such an odd number? Well, that is the gauge of English railways, and US railroads were designed by British engineers. Why did the English use such an odd number? Railroads were preceded by horse drawn trams, which had the same gauge. Builders of trams used the same jigs that they used to build wagons and the wheel width of wagons matched the ruts on the old stone roads built by the Romans. To do otherwise would have resulted in a lot of broken wheels. The Romans, who introduced the first bureaucracy into England, had a specification for war chariots that accommodated two horses.

Thus, a 2,000 year old specification from Imperial Rome, based on the width of a horse's behind, dictated design limits for the most advanced vehicle in the 20th century.

Chapter 23   The History Lesson in Terminology

Often, the terminology we use provides a hint about something from the past that is no longer relevant. Many of the terms used in aviation have maritime origins. For example, why do we say 'port' and 'starboard' instead of left and right?

The term 'starboard' dates back to the days before boats had a rudder at the stern. Many early sailors never left sight of shore. They would simply beach themselves through surf each evening, and then back out through the surf the following morning. Accordingly, a rudder at the stern would be vulnerable in the same way that the Wright Flyer's elevator at the front was vulnerable. As a consequence, early ships were steered by a steering board on the right side near the stern. Why the right side? That probably goes back all the way to when the earliest boats were paddled. Most people were right handed, so the boat would tend to veer to the left. This tendency was counteracted by a paddle or steering board on the right side of the boat. The farther back it was, the more leverage it got per the Lever Principle and the smaller the steering board had to be. Smaller meant less total drag, an approach that is also employed with airplane control surfaces.

The term 'port' meaning 'left side' is of more recent vintage. When the boat was docked at a pier, the steering board would be on the side away from the pier to avoid being crushed, so the side towards the pier would be 'the port side'. Originally the left side of a boat was referred to as 'larboard', probably derived from 'ladder board'. The ladder would be on the side towards the port and away from the steering board. Larboard and starboard sound so much like each other, that the term larboard was dropped and replaced with 'port'.

What is the origin of the term 'posh' which means elegant and expensive. Although it does not sound particularly nautical, posh stands for PORT OUT STARBOARD HOME and referred to the side of the ship that had the most desirable cabins when traveling from England to India and back home to England in the days when Britain ruled India. When crossing the Arabian Sea and the Mediterranean Sea one would pay more for a cabin on the side of the ship that was away from the Sun.

Is this just interesting trivia, or is there value in tracking down the origin of the terms we use? I have always found that it helped me to 'think outside the box" if I knew what defined the box in the first place. Clearly, knowing the shape of the box helps.

Steering Wheels – Left or Right? – Cause & Effect

Another paradigm that has ancient origins relates to steering wheels. Why is the steering wheel on a car on the left side in America and the steering wheel on a motorboat on the right side?

Go back in history and you will find that people traveled on the left side of the road when gentlemen wore swords. Most people were right handed and that put one's sword-hand towards any oncoming riders. After the French revolution the nobility and their trappings were out of favor, so the French switched to the right side (If you did not have a sword it would make sense to put yourself on the side where the oncoming person had the least advantage). Then came Napoleon who took over most of Europe and spread the French way of doing things across the Continent. However, the British never changed and drive on the left side of the road to this day.

The USA was still at war with Britain when the French Revolution began, so possibly we simply aped the French. Another possibility was the teamsters. When driving a team of horses pulling a cart that did not have a seat, and with a whip in his right hand, it made sense for the teamster to ride the last horse on the left side. Thus he would want any oncoming carts to be on his left so that he could best judge the distance between them. Whatever the reason, traffic in America wound up on the right side of the road and the driver on the left side of his vehicle. But why is the steering wheel on a motorboat always on the right side?

The answer lies in the direction in which the propeller turns, which is clockwise. Just as the torque of an airplane's propeller pushes the plane into a left roll, the propeller on a boat causes the right side of the boat to rise out of the water. Therefore – to minimize that tendency - it makes sense to put the one seat that you can be sure will be occupied on the right side of the boat.

My goodness - my Guinness!

"What is going on with your Guinness?" queried the voice that could be heard above the din in an Irish bar in Chicago.

"What are you talking about, boyo?" asked a bartender, a recent immigrant from County Kerry.

"The bubbles are all going down, and that's impossible!" was the reply. *"Archimedes' Law says that the bubbles of gas should go up and the liquid go down."*

"Sure Archie is long dead and gone, and that Guinness is doing what Guinness always does. Wait a minute if you don't want to see the bubbles going down" advised the bartender and went away to serve another customer.

I gave up and went back to my Guinness. It was usually easier to get a reaction with that question. In fact I had been using it for several years when conducting job interviews with applicants who had PhDs in chemistry. I found that it was a good way to sort out who could handle confusing data. I was looking for scientists who knew their stuff, but who could dissect a problem that defied logic. Out of 33 interviews, I only found one who nailed the solution, and two others who were on the right track. About a third of the job candidates dismissed the question out of hand. Their candidacy ended with that interview. I was looking for people who could work and think outside the box.

If you have never had a pint of Guinness, then you are forgiven for not knowing that you see the gas bubbles going down when the bartender puts the glass in front of you. That assumes of course that it was a properly served pint. Could the gas bubbles (CO_2) really be heavier than the liquid (mostly H_2O with a lot of dissolved solids)? Does Guinness defy Archimedes' Law?

The answer is of course "No", but Archimedes' Law is not the only science in play in this situation. For starters, while the bubbles you see are actually going down, you can only see a tiny fraction of the bubbles because of the opacity of the liquid. In fact, most of the bubbles are going up, but you can't see them. Not only are they going up, but they are pushing liquid ahead of them, which is why it takes so long to pour a perfect pint. It is actually a two step process as mentioned in the chapter on Archimedes' Law.

With all this liquid being pushed up, it has to go somewhere, and gravity says "down". However, the downward moving liquid is heavier than the gas and can push gas bubbles down, and guess where they do so. At the edges of the liquid where it is in contact with the glass surface, and these are the only bubbles that you can see. The gas bubbles that have only a gas/liquid interface meet less resistance than gas bubbles that have a gas/glass interface, and therefore the bubbles not in contact with the glass can move faster. Bottom line, Archimedes' Law dominates in the center of the glass, Newton's 2nd Law dominates at the edges, and Reynold's Number explains why there is a difference between the two areas.

Is this question idiotic, or was it a useful exercise? The person who nailed the answer was a Physical Chemist. He blinded me with a lot of physical equations about friction, inertial vs. viscous forces, etc. As an Organic Chemist, this is not an area of strength for me, so a lot of it went over my head. But that was ok. I wanted people who were smarter than me and who could apply their knowledge of science at the drop of a hat. I offered him a job. The other two who were on the right track also got job offers. To be good at science, you do not have to know everything, you just have to know where it is, who knows it, or how to find it. The confidence to use Cause & Effect analysis is especially important.

Another example of Cause & Effect analysis that involves Guinness came up when I was involved in a discussion of the declining "Pub Culture" in Ireland. I spend 2-3 months there every year. It was late at night and the few people left in the pub shared one attribute, grey hair.

> *"Why are there no young people here?"* was asked by one person. He meant people in their twenties, not teenagers.

That is the effect, what was the cause?

> *"The law banning smoking in bars and the strict limits on blood alcohol for drivers are the main reason"* said another individual. *"Young fellahs can smoke and drink in their own homes"*.

We all looked up at the sign "Michael Martín Lounge" over the side door leading into an alley alongside the pub. Michael Martin was the Minister of Health when the smoking ban on working places was introduced in Ireland. Most bars in Ireland now have an outdoor area where patrons who need a cigarette go out for a smoke. The maximum blood alcohol allowed for drivers is tighter in Ireland than in America, which has resulted in a massive increase in the number of taxis in small Irish towns. (Cause & Effect!)

> *"There is another reason",* said another. *"Anybody can pour a beer, and can do it anywhere. It takes skill, equipment, and patience to pour a Guinness"*.

The proprietress of the bar, a 3rd generation publican, nodded in agreement. Pouring a perfect Guinness is a two-step process involving specialized equipment and definitely is a more complicated process than pouring a beer.

> *"That is why most of the people who come to pubs today are Guinness drinkers. Beer and lager drinkers can get it in cans or bottles and don't need to come to the pub",* continued the knowledgeable one.

It is all Cause & Effect.

Chapter 24 STEM Q&A– Jargon & Terminology

Questions	Answers
"Mayday" is a signal sent out by a pilot who is about to crash. What is its origin?	"M'Aidez," French for "help me" is pronounced "Mayday" asks you to drop everything and help
"S.O.S" is the maritime equivalent of "Mayday". What is its origin and meaning?	"Save our souls" is often given as the meaning, but it really is derived from the simplicity of its Morse code signal: Dit-Dit-Dit Dah-Dah-Dah Dit-Dit-Dit
"Pan Pan" is sent by a pilot who is in trouble but not (yet) going down. Origin?	Literally "Possible Assistance Needed" is an alert to the receiver that help may be needed.
What does it mean when a pilot is said to have "bought the farm". Origin of the term?	Families of airmen killed in WW1 got stipends that were enough to buy a small mid-west farm.
What is happening when a pilot is "pushing the envelope"?	With airplanes the term "envelope" refers to a range within which it is safe to fly. e.g. The weight and balance envelope for the center-of-gravity or the performance envelope for speed. "Pushing the envelope" means going to or beyond safe limits.
The term "knots" means "nautical miles per hour". What is the origin of the term "knots"?	Ship speed was measured by throwing overboard a log tied to a rope with knots at regular intervals. They would count the number of knots running out in a fixed amount of time and record speed in the "log book".
What is the origin of the terms "Port" & "Starboard"?	Early boats that were pulled ashore through surf did not have a rudder at the stern. A "steering board" (starboard with the right accent) was near the stern on the right side. Alongside a pier, the steering board would be away from the dock. Hence the left side was the 'Port' side.
What is meant by the phrase "balls to the wall"? What is the origin of the phrase? **(Note: The answers are not rude)**	It was common for airplane throttles to be a knob in the shape of a ball. Thus, "balls to the wall" meant the throttle was pushed all the way forward. i.e. full speed. The throttle of an airplane performs the same function as an accelerator in a car.
The term "posh" means wealthy or high class and has both naval and weather related origins. What are they?	Port-Out-Starboard-Home referred to a cabin on the expensive but cool side of a ship from England to India across the Mediterranean and Arabian Seas.
What is a pilot doing when he is flying "by the seat of his pants"?	"Flying by the seat of one's pants" means using sensory references rather than instruments, a recipe for disaster under instrument conditions
What does it mean when a pilot says "Roger, wilco"?	Morse code for R (Dit-Dah-Dit) signified that a message had been received. In voice radio "Roger" meant "message received & understood". "Wilco" means "will comply" and is used for clarity, since "OK" could mean "I heard you" or "I'll do it".

Chapter 25

Doing it Digitally

We are in the Digital Age, so we should be doing things digitally. Right? But what does that mean? It means using computers. If you use Intuit's *Turbo Tax* to do your taxes, then you are doing them digitally.

Personally, I can't read computer manuals. However, I can usually adapt an example to meet my needs, and that is why I have the program shown below tacked to the wall by my desk.

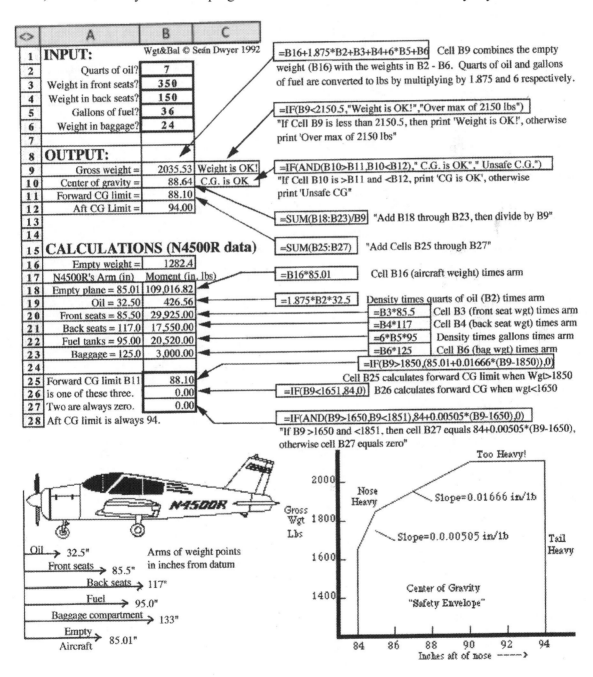

	A	B	C
1	**INPUT:**	Wgt&Bal © Seán Dwyer 1992	
2	Quarts of oil?	7	
3	Weight in front seats?	350	
4	Weight in back seats?	150	
5	Gallons of fuel?	36	
6	Weight in baggage?	24	
7			
8	**OUTPUT:**		
9	Gross weight =	2035.53	Weight is OK!
10	Center of gravity =	88.64	C.G. is OK
11	Forward CG limit =	88.10	
12	Aft CG Limit =	94.00	
13			
14			
15	**CALCULATIONS (N4500R data)**		
16	Empty weight =	1282.4	
17	N4500R's Arm (in)	Moment (in. lbs)	
18	Empty plane = 85.01	109,016.82	
19	Oil = 32.50	426.56	
20	Front seats = 85.50	29,925.00	
21	Back seats = 117.0	17,550.00	
22	Fuel tanks = 95.00	20,520.00	
23	Baggage = 125.0	3,000.00	
24			
25	Forward CG limit B11	88.10	
26	is one of these three.	0.00	
27	Two are always zero.	0.00	
28	Aft CG limit is always 94.		

=B16+1.875*B2+B3+B4+6*B5+B6 Cell B9 combines the empty weight (B16) with the weights in B2 - B6. Quarts of oil and gallons of fuel are converted to lbs by multiplying by 1.875 and 6 respectively.

=IF(B9<2150.5,"Weight is OK!","Over max of 2150 lbs")
"If Cell B9 is less than 2150.5, then print 'Weight is OK!', otherwise print 'Over max of 2150 lbs"

=IF(AND(B10>B11,B10<B12)," C.G. is OK"," Unsafe C.G.")
"If Cell B10 is >B11 and <B12, print 'CG is OK', otherwise print 'Unsafe CG"

=SUM(B18:B23)/B9 "Add B18 through B23, then divide by B9"

=SUM(B25:B27) "Add Cells B25 through B27"

=B16*85.01 Cell B16 (aircraft weight) times arm

=1.875*B2*32.5 Density times quarts of oil (B2) times arm

=B3*85.5 Cell B3 (front seat wgt) times arm

=B4*117 Cell B4 (back seat wgt) times arm

=6*B5*95 Density times gallons times arm

=B6*125 Cell B6 (bag wgt) times arm

=IF(B9>1850,(85.01+0.01666*(B9-1850)),0) Cell B25 calculates forward CG limit when Wgt>1850

=IF(B9<1651,84,0) B26 calculates forward CG when wgt<1650

=IF(AND(B9>1650,B9<1851),84+0.00505*(B9-1650),0)
"If B9 >1650 and <1851, then cell B27 equals 84+0.00505*(B9-1650), otherwise cell B27 equals zero"

Oil → 32.5"
Front seats → 85.5"
Back seats → 117"
Fuel → 95.0"
Baggage compartment → 133"
Empty Aircraft → 85.01"

Arms of weight points in inches from datum

Too Heavy!
Nose Heavy
Slope=0.01666 in/lb
Slope=0.0.00505 in/lb
Tail Heavy
Gross Wgt Lbs
2000
1800
1600
1400
Center of Gravity "Safety Envelope"
84   86   88   90   92   94
Inches aft of nose ----->

The spreadsheet contains English language translations of most equations that I need in an Excel program. For example, an *"If . . . and . . . then . . . or else"* type equation is shown in Cell C10. Read the translation. It is easier than explaining a two-condition equation in Geek speak. If you are a purist that does not consider writing Excel to be "programming", then you are free to do it the hard way using Fortran, Cobol, or some other computer language.

Step by Step

If the above program looks intimidating to you, be assured that it would have been so to me not too many years ago. The key to conquering the intimidation is to build a program in small steps.

For example, if you can type just seven measly lines of text you can calculate the density altitude at an airport. Three more lines can provide the takeoff roll of a store bought airplane like a Piper Cherokee. Three lines beyond that can introduce the impact of humidity. If you have gone that far, you might as well write a program that will calculate the performance of an aircraft that you build yourself. OK, I never built an airplane. But I did write the program. Let us take it in steps (and pay attention to the STEM issues involved).

Step 1 Calculating Density Altitude

'Density altitude' is the effective altitude of a given air density, and air density depends on temperature and the local atmospheric pressure. Pilots at high elevation airfields or short runways ignore it at their peril. You will understand as we go along. The first step requires just 7 lines in an Excel spreadsheet. Line B7 is a real humdinger. Don't let it intimidate you! All you have to do is type it in correctly. Your tax dollars already paid some government geek to figure it out. Trust him. The ensure that you typed it correctly, enter the same inputs in B2, B3, and B4 as shown. If you get the same output, then your typing was fine.Start Excel and enter the cells as shown below:

A2 Field elevation in feet msl?
A3 Airport altimeter setting?
A4 Temperature in degrees F?
A7 Density Altitude =
B1 Input
B6 Output
B7 $=145457*(1-(1.95467/(B4+459.7)^0.23494)*(0.01+(B3^0.19025-0.000013125*B2)^5.2564)^0.23494)$

To calculate density altitude, all you have to do is enter the current field elevation, altimeter setting, and temperature (the "Inputs") into Cells B2-B4, and the density altitude will appear in Cell B7. This is how it will appear on the screen:

◇	A	B
1		Input
2	Field elevation in feet msl?	683
3	Airport altimeter setting?	29.92
4	Temperature in degrees F?	94
5		
6		Output
7	Density altitude?	3,032

Note that an airfield with an elevation of 683 feet above sea level has an effective altitude of 3,032 feet. This is due to the high temperature. In accordance with Charles' Law (*An ideal gas will expand in direct proportion to its absolute temperature)*, the higher temperature caused the molecules of air in the atmosphere to spread apart and the resulting air density is what you would have expected to find several thousand feet higher in altitude. This will impact both the climb rate of the airplane and the length of the takeoff roll. Those will be calculated in Step 2.

Step 2 Density Altitude Vs. Take-off roll

◇	A	B	C
1		Input	
2	Field elevation in feet msl?	683	
3	Airport altimeter setting?	29.82	
4	Temperature in degrees F?	94	
5			
6		Output	
7	Density Altitude =	3,032	
8	Climb rate in feet/min =	483	(Default if zero)
9	Take-off roll @ 2150 lbs. =	1,185	1,185
10	(Default used if DA , 0 or > 7,000')		

Addition of rows 8,9 and 10 and use of the performance charts of a Piper Cherokee will determine its climb rate and take-off roll at various density altitudes.

B8 =IF(B7<0,630,630-0.04846*B7)
B9 =IF(B7>0,C9,800)
C9 =IF(B7<7001,800+0.127*B7,0)

B8 means "If Cell B7 (i.e. Density Altitude) is less than zero then B8 equals 630; Otherwise it equals (630-0.04846 times B7)."
B9 means "If Cell B7 is greater than zero, then Cell B9= Cell C9, otherwise Cell B9 = 800."
C9 means "If Cell B7 is less than 7001 feet, then C9 equals (800+0.127 times B7); Otherwise C9 equals zero."
These "If...then..." statements provide default values when density altitude is either less than zero or

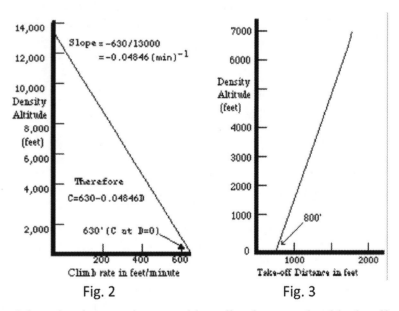

Fig. 2   Fig. 3

outside a Cherokee's performance charts. Otherwise the equations would predict that you should takeoff backwards at really cold temperatures (an unlikely event), or predict takeoff rolls that are not based on the performance charts at density altitudes >7000 feet msl (mean sea level). Refer to the performance charts in Fig. 2 and 3 to understand the basis for the numbers in the equations.

C8 (Default if zero) is just an explanation that a default number is used if density altitude is <0.

Step 3 Density Altitude and Humidity Vs. Take-off Roll

Does higher humidity increase or decrease density altitude? The surprising answer is that humid air is lighter than dry air. That is counter-intuitive for most people. The molecular weight of water (H_2O) is 18, whereas those of N_2, O_2 and CO_2 are 28, 32 and 44 respectively.

Avogadro's Law says *Equal volumes of ideal gases at the same temperature and pressure have the same number of molecules.* So any increase in the number of water molecules means a decrease in the number of other air molecules, and they are all heavier than water.

◇	A	B	C
1		Input	
2	Field elevation in feet msl?	683	
3	Airport altimeter setting?	29.92	
4	Temperature in degrees F?	94	
5	Dew point in degrees F?	85	
6			
7		Output	
8	Density Altitude =	3,032	
9	Temo increase due to humidity =	8	
10	Humidity corrected density altitude =	3,524	
11	Climb rate in feet/min =	483	(Default if zero)
12	Take-off roll @ 2,150 lbs =	1,248	1,240
13	(Default used if DA < 0' or > 7000')		

To account for the effect of humidity we need to add the dew point, which is part of any weather report from an airport, and two formulas to calculate the effective temperature increase due to humidity and then correct density altitude for this new "effective temperature".

Insert a new row at row 5 and enter the following:
A5   Dew point in degrees F=

Insert two new rows after the Density altitude row (which would now be row 8, having just been moved down). Enter the following:
A9   Temperature increase due to humidity =
A10 Humidity corrected density altitude =
B9   =IF(B5<31,0,2.88315-0.1267*B5+0.00223*(B5)^2)
B10 =145457*(1-(1.95467/(B4+B9+459.7)^0.23494)*(0.01+(B3^0.19025-0.000013125*B2)^5.2564)^0.23494)
B11 =IF(B8<0,630,630-0.04846*B8)
B12 =IF(B8>0,C12,800)
C12 =IF(B10<7001,800+0.127*B10,0)

To use the program, enter field elevation, altimeter setting, temperature and dew point into cells B2-B5 and the spreadsheet does the rest. Those inputs are all part of a standard weather briefing for a pilot. When would a pilot need to do this type of calculation? The pilot is *always* required to know that he has enough runway to take off. So, if you have a short runway and it is a hot day, do the calculation. If you are at a high elevation airport and it is a hot day, then the density altitude may be beyond the capability of your airplane, which is why mountain flyers often take off early in the morning or in the evening when it is cooler.

Factory built aircraft come with detailed charts that show climb rates, takeoff distances, etc. Of course, those charts reflect the aircraft's performance when it was new, and not what you now get forty years later. My Piper Cherokee is a 1965 model, and it is still going strong. However, I do not kid myself that it is as good as new and wrote an Excel program to generate a performance chart. This comes next. An airplane home-builder might also find it interesting.

Creating Performance Charts

In the program that follows, one selects an airspeed to evaluate and then records the climb rate, outside air temperature, and pressure altitude while climbing at this airspeed. I selected 83 mph, in the example below. Pressure altitude is determined by setting the altimeter to 29.92". Data points are taken at convenient intervals. Although all the readings for a particular speed can be gathered on a single flight, data from different flights can be combined as long as the aircraft is the same weight each time and the same indicated airspeed is used.

Air Speed =	83	mph	
Pressure Altitude	Air Temp	Climb Rate	Density Altitude
2.000	42	1,100	1,342
3,000	43	950	2,649
4,000	43	750	3,886
5,000	40	700	4,923
6,000	37	600	5,960
7,000	34	500	6,998
8,000	30	450	7,969
			(4,185)
			(4,185)
			(4,185)
			(4,185)

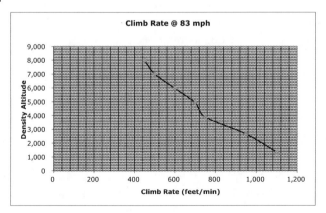

Enter the program as follows:
Cell A1   Air Speed =
Cell C1   mph
Cell A3   Climb Rate
Cell B3   Pressure Altitude
Cell C3   Air Temp
Cell D3   Density Altitude
Cell D4 $= 145300*(1-(518.4/(B4+459.4)*(1-A4/145300)^{5.255})^{0.235})$

Repeat the equation in Cell D4 in Cells D5 through D 14 using the FILL DOWN function under the EDIT menu. This will automatically advance the B4 to B5 and A4 to A5 etc. If the speed used is 83 mph, then save the program as "83mph.xls".

To chart the data, highlight the cells with the climb rate and density altitude data – including the headings in row 3 - and select CHART under the INSERT menu. Then select "XY (Scatter)" under 'Chart Type' followed by the option with data points and curves under 'Chart sub-type'. Click on "NEXT" and then "Next" again to get a page that includes a bar with "Title, Axes, Gridlines, Legends, and Data Labels" across the top. Select "Titles" and then enter "Climb Rate at 83 mph" in the Chart Title box, followed by "Climb Rate (feet/min)" in the "Value X Axis" box, and "Density Altitude" in the "Value Y Axis" box.

Click on "Gridlines" and check the "Minor gridlines" boxes for both axes. Click on "Legend" and uncheck the "Show legend" box and you will get the graph shown above.

What airplane was I in when I generated the data shown above? Probably my own Cherokee, as it is one of the few planes around that show speed in mph only and not nautical miles. However, I was clearly nowhere near the maximum gross weight at the time.

Spreadsheets

I vividly recall the first time I saw a spreadsheet program demonstrated. Managers in my division were doing budgets, and we all saw the time savings it promised. Most became believers and learned to use the program. However, too many people have not made the transition from when writing a program was strictly the province of geeks. Excel is not like writing Fortran. It is a skill set that everybody should develop, particularly since every computer in most corporations has the Microsoft Office suite of programs, and that includes Excel.

One of my pet peeves was when someone would buy a specialized program to do something that Excel could have handled easily, and then expected others to work with their output. Frequently, the result would be that just three or four people in the company knew how to use the unique program. If the data had been generated on Excel, every computer in the company could have worked with it, and people would reinforce each other's skill sets. Perhaps more importantly, data sharing across different fields and functions gets people to think "outside the box".

Undoubtedly, some departments use unique programs to maintain control over an activity, and operate as 'functional silos'. (silo = sealed chamber designed to protect its contents) Once when I was working overseas, my boss asked, "Did you give the Swedish Marketing people a P&L program? The Finance people do not think an R&D manager should do this".

"Well why don't the Finance people figure out what are our real cost generators?" was my reply. It was in the early days of spreadsheet programs and I was uncovering opportunities that will be discussed in the next chapter, STEM Vs. Paradigms. The Digital Age is a real eye-opener.

Skill sets are enabling

Everybody should take on a personal goal to learn a new skill set every year. Keep at it until people start asking you for advice in that skill. Then you know that it is time to find a new skill to learn.

Two skill sets that I had pursued were Microsoft® Excel and learning to fly. As I was a development chemist who later went into R&D management, the flying part did not appear relevant to my job. However, the two unrelated skills merged when someone asked me to find a solution to collapsed blister packs that were arriving in our distribution center. The first issue to address was why they were collapsing in the first place. Was the product inside shrinking? Was some component leaching out through the wall of the pack? Nobody knew.

I was aware that the packages were being sealed at a plant in Mexico. After obtaining aviation weather reports from a nearby airport in Mexico I wrote an Excel program to calculate the air density when the packages were sealed and when they arrived at the distribution center. The logic was similar to determining the performance of an airplane at different density altitudes.

The atmospheric pressure difference amounted to two miles of altitude! Our options were either to incorporate an expansion/contraction panel in the blister pack or seal it in a factory at a lower elevation.

Chapter 26   Dead Reckoning at Midway

June 4, 1942

"Dead Reckoning, I sure don't like the sound of that", thought the young pilot. It was more than an hour since VB-6 (Bomber Squadron 6) had launched from the USS Enterprise on that fateful day in June 1942. They were searching for the Japanese fleet that was believed to be approaching Midway Atoll. Not only would they have to attack that fleet, but then they would have to find their way back to their own carrier, depending primarily on 'dead reckoning'.

To make the math of navigation a little more exciting, this chapter on dead reckoning uses a historical context. Imagine that you are a pilot at the Battle of Midway, and you need to navigate back to your aircraft carrier across trackless ocean. If you run out of gas before you find the carrier or a friendly ship, you will have to ditch in the ocean and hope that somebody finds you before the sharks or thirst kill you.

This chapter could just as easily have been titled "Navigation by the Ancients", as the reality is that carrier pilots flying across trackless ocean in World War 2 used the same dead reckoning that Columbus used, as did Phoenician traders a thousand years before him. Even though the fighters and bombers had radios, they could not call their aircraft carriers for a 'fix', as that would have revealed the position of the carriers to the enemy. This was something they could not afford to do less than six months after the carnage at Pearl Harbor.

As a form of navigation, dead reckoning tells you where you are relative to a starting point, even if you do not know where that starting point is relative to the rest of the world. That worked for carrier pilots who were out in the middle of nowhere. However, we will get back to that later. A quick history of the war to that point is in order to get you into the right frame of mind. Navy thinking at the time was a good example of paradigm paralysis. Too few of the admirals on the American side were aware that the era of the battle ships was coming to a close. Aircraft carriers were about to emerge as the key weapons at sea.

Pearl Harbor December 7, 1941

It was a Sunday morning. Undetected by the US Navy, a Japanese fleet that included six large aircraft carriers (*Akagi, Kaga, Soryu, Hiryu, Shokaku,* and *Zuikaku)* had approached to within 200 miles of Pearl Harbor and launched a surprise attack.

Additional Japanese carriers available at the time included *Shoho, Zuiho, Hosho, Junyo,* and *Ryujo,* although they were not involved in the Pearl Harbor attack. In contrast, the USA had only three carriers in the Pacific at the time, *Lexington, Enterprise,* and *Saratoga.* Already at a 4 to 1 disadvantage in carrier numbers, the USA was particularly fortunate that none of its carriers was in Pearl Harbor on December 7th. *Enterprise* and *Lexington* had been sent to deliver aircraft to Wake Island and Midway Atoll respectively, and *Saratoga* was entering San Diego harbor to pick up her air group, which had been training ashore while *Saratoga* was being refitted.

That was not the only good luck on the part of the US Navy that day. The Japanese pilots were so intent on attacking the eight battleships, that they did not destroy either the fuel storage tanks on top of a nearby hill or the dry docks in the harbor. A single fighter could have strafed the fuel tanks and destroyed much of America's fuel in the Pacific theater. While all eight battleships were hit, and four of them sunk, two of the latter were raised and six of the eight battleships returned to service in the war. The availability of the dry docks in the same harbor played an important role in that.

Doolittle Strikes Back

With the exception of the Doolittle Raid on April 18, 1942, everything went in Japan's favor for the next six months. Militarily, the Doolittle Raid did negligible damage, although its psychological impact was considerable in both Japan and the USA. In a one-way attack on the Japanese homeland, Army B-25 bombers had taken off from the carrier USS *Hornet*, which had come through the Panama Canal in March. Because the B-25s could not land on the carriers, they had to continue on to China in the hope they could join up with friendly forces there.

Corregidor and the Philippines were conquered by May 6, 1942. The Battle of the Coral Sea (May 4-8, 1942) was the first major sea battle where none of the ships came within sight of those of the enemy. All the fighting was done by the air wings of each side. Japanese planes sank the USS *Lexington* and left the USS *Yorktown* in flames. *USS Yorktown* had also come west from the Atlantic following the attack on Pearl Harbor. While the Battle of the Coral Sea may have been a tactical victory for Japan, it was a strategic defeat, and their forward progress towards Australia was halted. Japan lost the light carrier *Shoho* and the fleet carrier *Shokaku* was severely damaged. In addition, the air complement of another fleet carrier *Zuikaku* was heavily depleted. The absence of these two veterans of the Pearl Harbor Attack was to prove decisive in the Battle of Midway just a month later. Now, back to our pilot's dead reckoning problem at Midway.

Dead Reckoning

Dead reckoning is the coordinated use of time, speed, and direction to determine one's location relative to some starting point. While Columbus sailed at 5 knots for 3 weeks, the pilots at Midway flew at 150 knots for 3 hours. The difference was in scale, not concept, and dead reckoning is simple applied geometry, complicated by the need to land before your fuel runs out. It was not like a test in high school, as your life really depended on getting the right answer.

Dead reckoning for a pilot has the additional complication of changes in wind direction. While the wind direction at sea level might be read by looking at the surface of the water, the speed of that wind would be purely a guess. Also, one does not fly at sea level. Whatever the wind speed at cruising altitude, that guess would have to be combined with airspeed and course(s) to estimate the aircraft's position relative to its home carrier. The plural, course<u>s</u>, is another complication, and pilots had to carefully note when they changed course and for how long they flew in any particular direction. Relying on the Flight Leader to get the flight back to the carrier would not be an option if the Flight Leader did not survive the attack.

Complicating the navigation problem even further was the knowledge that *Enterprise* would not have been sitting still on the water and could have moved 100 miles in the four hours they would be away. In case you think navigating back to the carrier was simply a matter of flying the reciprocal of your outgoing course, think again. For example, if you flew 270° (due west) from point A to Point B, using a course of 090° (due east) would not get you back to the carrier if the aircraft carrier was sailing 180° (due south) at 30 knots during the hours you were in the air.

Consider the example of Bomber Squadron 6 from *Enterprise* and you will see how complicated it gets. They flew 50 minutes on a course of 243°, followed by a course of 324° for 65 minutes, and then 053° for 35 minutes before they found Admiral Nagumo's fleet.

Assuming a cruise speed of 150 knots for a Douglas SBD-3 Dive Bomber, they would already have traveled 375 nautical miles on three different courses before they found the Japanese fleet. Fuel would already be a concern. Where were they relative to *Enterprise?*

"I wish I paid more attention in geometry class", went through the pilot's mind as he marked the second change in course on his lapboard. His Squadron Leader, Cdr McClusky, had just spotted a Japanese destroyer going somewhere in a hurry and the squadron turned to match its course. They had already made one course change to the northwest when they found empty ocean where the Japanese fleet had been reported.

Spotting the *Ararhi* was one of those quirks of fate where pure luck dictates the outcome of a battle. America won the Battle of Midway and it was a turning point in the war with Japan. By the time it ended, four Japanese aircraft carriers had been sunk and the USA lost the *Yorktown*. Japan never recovered from the loss of four carriers and their fate was hidden from the Japanese public for the rest of the war. Luck and courage played a huge role on both sides. The following synopsis of the battle demonstrates the confusion of war.

Code Breaking

America had broken the Japanese naval code and was expecting attacks on both Midway Atoll in mid-Pacific and the Aleutian Islands off the coast of Alaska that week. The carriers *Enterprise* and *Hornet* were already in ambush position before the Japanese submarine screen was put into place. *Yorktown,* which had been badly damaged in the Battle of the Coral Sea days earlier, was rushed through repairs and also got into position without being seen by the submarines. However, it was spotted by a Japanese scout plane.

The presence of three American carriers would be a complete surprise to the Japanese, as they believed the *Yorktown* had been sunk during the Battle of the Coral Sea along with the *Lexington*. They were expecting that *Enterprise* and *Hornet* would sail from Pearl Harbor after they attacked Midway, not before. Japan's goal for Midway was to lure the American carriers into a trap and complete what they had been trying to achieve in the attack on Pearl Harbor, a time when the American carrier fleet had been at sea.

Unlike the Americans, over-confidence on the part of the Japanese allowed them to divert two smaller carriers, *Ryujo* and *Junyo,* far to the north to attack the relatively unimportant Aleutian Islands. Furthermore, unlike the frenzied efforts to repair *Yorktown*, Japan's efforts to repair the fleet carrier *Shokaku* – which had also been damaged in the Battle of the Coral Sea - were such that it remained behind in Japan. Japan had six carriers for the Pearl Harbor attack, and only four for Midway, a crucial mistake for the Japanese.

Before our side learned the location of the enemy fleet, Adm. Nagumo initiated the attack on Midway's airfield. Meanwhile, the American submarine *Nautilus* attacked the Japanese fleet and Nagumo learned the position of the *Yorktown* from a scout plane. After recovering the aircraft that had attacked Midway he steamed northeast to close with *Yorktown,* leaving the destroyer *Ararhi* to drop depth charges on *Nautilus.*

Some American fighters and bombers (VF-8, VB-8 and VS-8 from *Hornet* arrived where the Japanese fleet was reported and turned south, expecting Nagumo to have done so to recover his planes from the Midway attack. They played no further role that day. Others turned north. The first to find the Japanese fleet was VT-8 from *Hornet.* Arriving with no fighter escort, the slow torpedo bombers were all shot down. Almost immediately after that, 14 torpedo bombers from *Enterprise* (VT-6) attacked and all but 4 were destroyed with no damage to the enemy. The Japanese were amazed at the number of aircraft that had attacked them, but felt lucky about their success in fighting them off with no damage to themselves. However, most of their fighter top cover had come down to fight off the torpedo bombers, who flew just above the water when they attacked.

Meanwhile, VB-6 (with the pilot in our exercise) turned northwest after finding empty ocean. It was only their sighting of the *Ararhi,* which had given up trying to sink *Nautilus* and was racing back to the fleet, that allowed them to find the fleet. They dive bombed the Japanese fleet which no longer had any air cover, and destroyed three of the four aircraft carriers, the *Soryu, Akagi,* and *Kaga.* Meanwhile, a Japanese attack had left the *Yorktown* burning and dead in the water, so the Japanese believed there now could be at most only one American carrier left. They launched another attack from Hiryu and again found the *Yorktown,* which had managed to put out the fires and was underway again. Once more they thought they destroyed it. The remaining Japanese carrier, *Hiryu,* was attacked and sunk, but this did not end threats to *Yorktown.* Because she appeared ready to capsize, the captain ordered *"Abandon Ship".*

    Yorktown was still afloat the following morning and Captain Buckminster went back on board with a salvage crew. The destroyer *Hammann* tied up alongside and provided power for pumps. By mid-afternoon things were going well. Then a Japanese submarine got into firing position and fired 4 torpedoes. One struck the *Hammann,* sinking her in minutes, and another hit *Yorktown.* As *Hammann* sank, depth charges on board reached their targeted depths and exploded. Still, *Yorktown* refused to sink. Salvage attempts were abandoned for the night. However, the following morning the list was even more pronounced and *Yorktown* soon capsized and sank in 3,000 fathoms of water.

    The Japanese Navy believed they had sunk *Yorktown* with aircraft on three occasions before a submarine finally did the job. Such is the confusion of war.

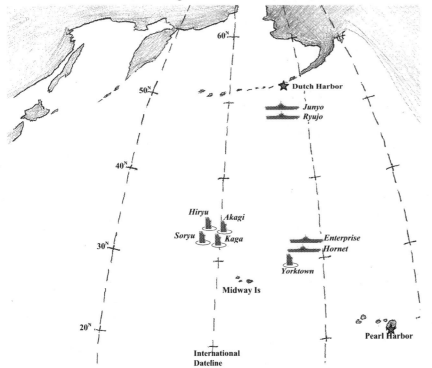

Back to our pilot from VB-6

    Now that you know the story, are you ready to do some Dead Reckoning and find your way back to the USS *Enterprise* before you run out of gas? You will need a naval pilot's plotting board to do so, and can find instructions on how to make one in Appendix 4. Do so, it is fun and the scenarios that follow are a great way to experience the confusion of war, without having anybody trying to kill you.

    The following page will show you how a pilot in VB-6 would use a naval pilot's plotting board. What it cannot do is replicate the tension, the fear, the concern for squadron mates, and even the injuries that would have been in play in the cockpit. They knew that they had left three Japanese carriers in flames. What would they find when they returned to their own fleet? Had it been attacked while they were away? Was their own carrier still afloat? My admiration for these men is unbounded. If you have not seen it, I recommend that you watch the movie **Midway** and read the book *Shattered Sword* by Jonathan Parshall and Anthony Tully.

    After reading through the instructions on how to use the plotting board, use it to answer the questions in the Midway Challenge. Some of the scenarios actually happened. Others are hypothetical, but even they would have been on the minds of the pilots on that fateful day.

Instructions for using the Navy Pilot's Plotting Board

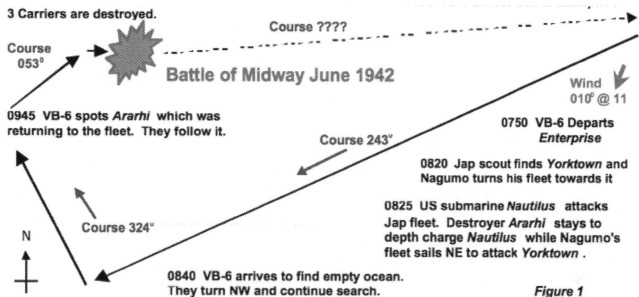

3 Carriers are destroyed.

Course ????

Course 053°

Battle of Midway June 1942

Wind 010° @ 11

0945 VB-6 spots *Ararhi* which was returning to the fleet. They follow it.

0750 VB-6 Departs *Enterprise*

Course 243°

0820 Jap scout finds *Yorktown* and Nagumo turns his fleet towards it

0825 US submarine *Nautilus* attacks Jap fleet. Destroyer *Ararhi* stays to depth charge *Nautilus* while Nagumo's fleet sails NE to attack *Yorktown*.

Course 324°

N

0840 VB-6 arrives to find empty ocean. They turn NW and continue search.

Figure 1

The plotter has a transparent rotating 360° compass circle on a 2-dimensional grid with parallel red lines that are 100 miles apart and blue parallel lines 10 miles apart. To plot distances and courses the rotating compass is aligned with one of the two red lines on the grid that run through the rotation point. When marking courses on the plotter use an erasable grease pencil. The course of VB-6 is displayed.

Time	Minutes	Course	Distance
0750-0840	50	243°	125 miles
0840-0945	65	324°	162.5
0945-1020	45	053°	112.5
0750-1020*	160	190°	29.3

*Wind correction for a speed 11 mph from 010°

At 150 knots, an SBD Dauntless flies 125 nautical miles in 50 minutes (150/60 x 50 = 125). Airspeed and times of course changes shown on the map of the battle provide the table on the right. The 4th course corrects for wind speed over the time from when VB-6 left *Enterprise* to when they engaged the enemy.

Plots are drawn on the circle.

Step 1: Rotate circle until the 243° radial overlaps the horizontal distance line. Draw line out to 125 miles to point Alpha ('A').

Step2: Rotate circle until the 324° radial overlaps the vertical distance line. All vertical lines on the grid are parallel and therefore at 324°. Starting at Alpha, draw a vertical line up 162.5 miles to point Bravo

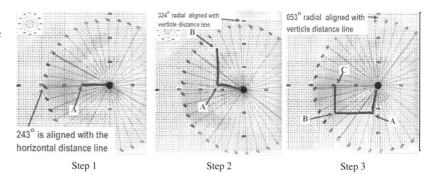

243° is aligned with the horizontal distance line

Step 1

324° radial aligned with verticle distance line

B

A

Step 2

053° radial aligned with verticle distance line

C

B

A

Step 3

Step 3: Rotate the circle until the 053° radial overlaps the vertical distance line. From Bravo draw a line vertically 112.5 miles to point Charlie ('C'). This is where VB-6 found the enemy.

Step 4: The wind correction covers 160 minutes that VB-6 was flying and assumes that wind was constant at 11 mph out of 010°. While unlikely, it is the best information available. Rotate the circle until the 190° radial overlaps the vertical distance line at the bottom of the plotter. Draw a 29.3 mile line parallel to the vertical distance line from Charlie to point Delta ('D').

Step 5: To determine bearing to point of departure, align 360° degree line on the circle with the vertical distance axis. Delta is approximately 311° <u>from</u> where they left the carrier, or 131° <u>to</u> it.

Step 6: Rotate the compass circle until Delta is on either distance measuring line. The distance to the departure point is the distance between Delta and the rotation point.

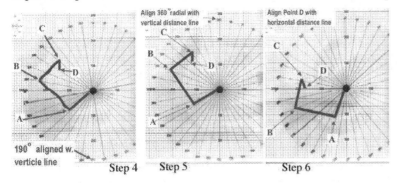

Midway Challenge (Answers can be found in Appendix 3)

Scenario 1: Petty Officer Nakoda (True)

You are Petty Officer Pilot Nakoda of the Japanese Imperial Navy. Your task will be to find your way back to the *Hiryu* after attacking and (you think) destroying *Yorktown*. We start when Adm. Yamaguchi on board the aircraft carrier *Hiryu* sees *Kaga, Akagi,* and *Soryu* all burning after the attack by American dive bombers. *Hiryu* had not been seen by the Americans. Yamaguchi tells Lt. Nicio Kobayashi that he wants him to find the American fleet. After being attacked by too many torpedo bombers to have all come from a single carrier, the Japanese commanders knew that they were up against at least two American carriers, possibly three. Adm. Nagumo sent scout planes looking for the American fleet, and Scout 4 radioed the position of the fleet 4 hours earlier, but did not mentioned any carriers. A fast scout was then sent out from *Soryu* to determine the types of ships. This plane actually found the three American carriers, but could not report this until he got back because of radio trouble. Yamaguchi sent Kobayashi with eighteen Val and six Zekes to the last position reported by Scout 4. ('Val' was the American name for the Aichi D3A fighter bomber which had a **cruise speed of 184 mph** and a maximum speed of 240 mph. 'Zeke' referred to the Mitsubishi Zero, an extremely maneuverable fighter with a long range. You (Petty Officer Nakoda) piloted a Val. You needed luck to find the enemy, but luck was on your side and now have to find your way back.

After flying 080° at cruise speed for 45 minutes, Kobayashi's radioman/gunner spotted SBD Dauntless dive bombers far below returning after having destroyed three of the four Japanese carriers. Kobayashi used light signals to order the Zekes to not attack. Just as American dive bombers had followed the destroyer *Arahi* to the Japanese fleet an hour earlier, the Japanese would now follow American planes back to *Yorktown*. **Their new course was 040° and they held it for 50 minutes** until they saw *Yorktown*. Kobayashi did not survive the attack, but you did. You radioed back *"Zephyr calling Whirlwind, Zephyr to Whirlwind. Enemy carrier ablaze. Returning to base."*

> **Your Challenge: What is your bearing and distance to *Hiryu*?**
> Note: **wind at take-off was 360° at 15 mph**.

Scenario 2: (Hypothetical!) U.S. Carriers Head South After Launching Airplanes

Through code breaking, the US Navy knew that two Japanese carriers *(Ryujo & Junyo)* were involved in the attack on Dutch Harbor in the Aleutian Islands on June 3. They had not been seen for a day and could have been steaming south-west towards Midway. They did not do so until it was too late for Japan, but the U.S. did not know this. It would not have been illogical for the American fleet to have told the pilots of VB-6 from the *Enterprise* that they would steam due south (i.e. **180° at 32 knots/mph**) once the planes launched. Zig-zagging to confuse enemy subs would reduce forward progress by about 20%, more if submarines were actually sighted. Your options include navigating back to where the carrier had been and then turning south, or using Wayne Gretzke's approach for great hockey *("Skate to where the puck is going to be")* to optimize fuel use. Both have advantages/disadvantages (What are they?) Assume that your course to the Japanese fleet were the actual courses taken by VB-6.

> **Your challenge: What course(s) should you fly to get to the new (hypothetical)**
> **location of *Enterprise*?**

Scenario 3: (Hypothetical) U.S. Planes have no choice but to fly to Midway Atoll

Assume that the attacks on the U.S. carriers involved all three American carriers and not just *Yorktown*. What if this attack was just as successful as the American attacks on the *Kaga, Akagi,* and *Soryu,* and the American had no carriers to which they could return? Their options could either ditch near a destroyer or try to get to Midway. When VB-6 departed from *Enterprise* the pilots knew that it was 200 miles NE of Midway on a bearing of 40°. 160 minutes later VB-6 participated in destroying the Japanese aircraft carriers. They hear on the radio that *Enterprise, Yorktown,* and *Hornet* had all been disabled, and their only options are to ditch near a destroyer or fly to Midway.

Your challenge: What would be their course and distance to Midway Atoll?

Scenario 4: How long would it take Jap carriers at Dutch Harbor to reach Midway fleets?

Kaga, Akagi, and *Soryu* were all crippled or sunk in the first attack on June 4, 1942, followed by *Hiryu* later that day. On the previous day Dutch Harbor in the Aleutian Islands had been attacked by a Japanese fleet that included two carriers, *Junyo and Ryujo.* This did not surprise Adm. Nimitz, as the decrypted messages revealed that an assault against the Alaskan islands would precede the Midway attack. However, were the two carriers that attacked Dutch Harbor a threat to the American carriers at Midway? The Japanese carriers in the Northern Fleet were approximately 1,700 nautical miles northeast of Midway on June 3rd, at a time when the *Enterprise* and *Hornet* were about 200 miles northeast of Midway. If they steamed at 35 knots towards Midway immediately after the attack on Dutch Harbor, could *Junyo* and/or *Ryujo* trap the American carriers between two powerful Japanese fleets? **What overall course would they use and how long would it take to be within attacking distance (i.e. c. 200 miles) of the American carriers?** To allow for zig-zaging to frustrate attack by subs, allow 20% extra sailing distance.

Your challenge: Determine what overall course the *Junyo* and *Ryujo* would use and how long it would take to be within attacking distance (i.e. within 200 miles) of the American carriers. You must zig-zag to frustrate attack by subs, so allow 20% extra sailing distance.

(Answers to all of the Midway Challenges can be found in Appendix 3)

Why Did *Chicago Tribune* Reveal that U.S. Broke Japanese Naval Code?

On Sunday, 7 June, the *Chicago Tribune* published its version of the battle under the headlines:

JAP FLEET SMASHED BY U.S.
2 CARRIERS SUNK AT MIDWAY
NAVY HAD WORD OF JAP PLAN TO STRIKE AT SEA
KNEW DUTCH HARBOR WAS A FEINT.

Somehow, the *Tribune* knew far more than the Navy had revealed. The story gave the names of 4 Japanese carriers, 2 battleships, 2 heavy cruisers, and 12 destroyers, seemingly lifted from a secret Navy document. Two headline phrases - Navy Had Word and Knew Dutch Harbor Was A Feint - suggested that the US Navy had some way of learning about Japanese operational plans.

While the *Tribune* story did not say the Navy had broken the Japanese naval codes, the suggestion was so strong that Secretary of the Navy Frank Knox, former publisher of the *Chicago Daily News,* immediately contacted the *Tribune* and other newspapers. Without telling the editors why, Knox asked them to halt further coverage of the story. Knox's action was not enough for Admiral Ernest J. King, commander-in-chief of the U.S. Fleet. Infuriated at the leak, he ordered an investigation. Pres. Roosevelt was so mad that he wanted to send Marines into the *Chicago Tribune* to arrest its publisher, Robert P. McCormick, for treason. He was talked out of it, for fear that a trial would tell the Japanese about the code breaking, if somehow they had missed reading it in thousands of newspapers.

The Leak

The carrier, USS *Lexington*, had been sunk on 8 May in the Battle of the Coral Sea and USS *Yorktown* was badly damaged. Among the survivors of the Lexington was Stanley Johnston, a *Tribune* war correspondent. Johnston and many other survivors were taken aboard the transport USS *Barnett,* which was bound for San Diego. He shared a cabin with his friend, Commander Morton T. Seligman, executive officer of the *Lexington.* At sea, Johnston read a decoded intelligence bulletin, transmitted on 31 May.

The information in the bulletin had been gleaned from the decoded intercepts that shaped Adm. Nimitz' strategy. Intelligence officers at Pearl Harbor had learned of the existence of a massive Japanese fleet and had deduced its probable objectives: Midway and the Aleutians. Analysis of the intercepts produced a recommendation about the best place to assemble an attack. All of this work was now at least indirectly revealed by the *Tribune* story.

Nimitz sent his three carriers—the *Hornet, Enterprise,* and the quickly repaired *Yorktown*—to a rendezvous point for an attack on the Japanese fleet. As expected, the Japanese aircraft attacked Midway, but did not destroy the airfield. Meanwhile, a Japanese search plane reported "what appears to be ten enemy surface ships" about 200 miles from the Japanese carriers. At almost the same moment, the *Hornet* and the *Enterprise* began to launch their aircraft toward what they believed to be two Japanese carriers. The *Hornet's* 15-plane Torpedo Squadron Eight flew into a swarm of Zeros and an inferno of antiaircraft fire. Japanese defenders wiped out the squadron.

While Zeros and antiaircraft crews concentrated on the torpedo planes, Douglas SBD Dauntless dive bombers from the *Enterprise,* far overhead, went into steep dives over two Japanese carriers, the *Akagi* and *Kaga.* SBDs from the Yorktown dived on a third, the *Soryu.* Within minutes, the three Japanese carriers were aflame and doomed. The fourth carrier, the *Hiryu,* had only hours to live. She, too, would be destroyed by dive bombers, but she had already launched dive bombers that had mortally damaged the *Yorktown.*

Roosevelt Vs McCormick

President Roosevelt and McCormick 'had history' with one another, McCormick despised Roosevelt's 'New Deal', believing that big government and wealth redistribution were socialist policies and not allowed by the Constitution. If you do not see the correlation between this and today's interaction between Obamacare, the Stimulus, the UAW bailout and the Tea Party, then you need to pay more attention to the news.

Remember, if you do not learn from history, you are doomed to repeat it!

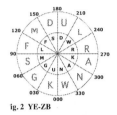

ig. 2 YE-ZB

- - - -TOP SECRET - - - -

Until they knew for sure that their position had been located by the enemy, aircraft carriers would maintain radio silence and not communicate directly with their aircraft. However, the U.S. had a secret called YE-ZB.

YE-ZB (Fig. 2) ('Zed Baker' to Navy pilots) was a weak UHF "line of sight" radio signal sent out from an aircraft carrier. The frequency could not be heard over the horizon, meaning that a ship beyond the horizon could not pick it up whereas an aircraft with sufficient altitude might.

The Morse code signal would be different in each of twelve 30° pie segments. Although land based codes never changed, codes used by aircraft carriers changed each day and would be recorded (in red) by pilots on their lapboards before departing. Note the compass points are opposite to what you normally see. This is because they represent direction from the receiving airplane **to** the carrier, and not its position **from** it. Thus a pilot southwest (i.e. 225°) of the carrier would steer 045° to arrive at the carrier. He would hear Morse Code for "G" (- - •) and know which pie piece he is in at that moment. If he flew due north the signal would eventually change to "S" (•••). At that point he would know that a course of 060° would take him directly to the carrier.

124

Appendix 1

US Patent # 43,449 to Solomon Andrews 1864
(See http://www.freepatentsonline.com/0043449.pdf for the complete text)

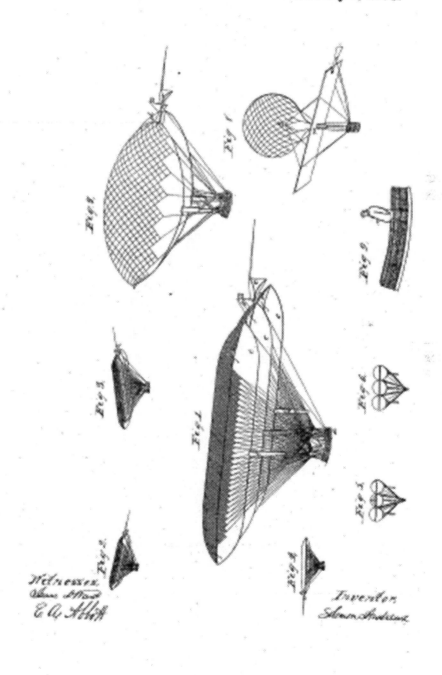

S. ANDREWS.
AEROSTAT.

No. 43,449.

Patented July 5, 1864.

Appendix 2

Dec. 30, 1969   J. R. FITZPATRICK ET AL   3,486,719

AIRSHIP

Filed Dec. 4, 1967   4 Sheets-Sheet 1

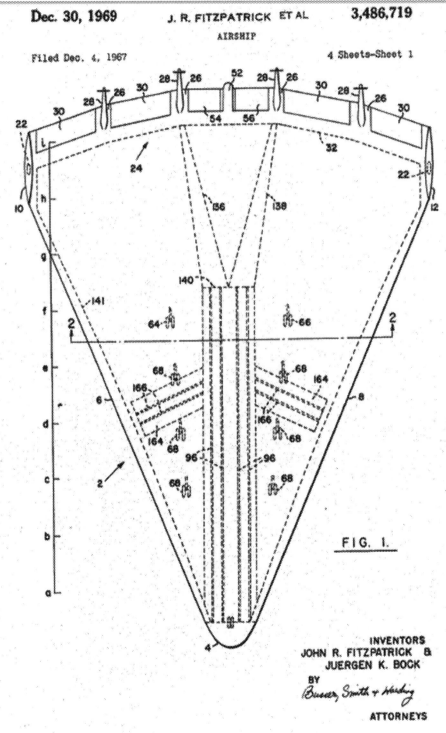

FIG. 1.

INVENTORS
JOHN R. FITZPATRICK &
JUERGEN K. BOCK

BY

ATTORNEYS

126

Appendix 3
Midway Challenge Answers

Scenario 1 Petty Officer Nakoda
The data provided says that he was cruising at 184 mph and wind when he left *Hiryu* was 360° @ 15 mph.
He flew: 080° for 45 minutes = 138 miles
040° for 50 minutes = 153.3 miles
180° for 95 minutes @ 15 mph = 24 miles
Petty Officer Nakoda was 63° and 255 miles from where he left *Hiryu*.
So his course to *Hiryu* was 243°. When he arrived and found *Hiryu* in flames, he had no choice but to land in the water near a destroyer.

Scenario 2 U.S. Carriers Sailed South After Launching Planes
When you turn back to your carrier the distance is 230 miles and the bearing is 141° **To** the carrier. Alternatively, you could go back to where you took off (180 miles @ 153° **To**) and then go south. The carrier will already be 80 miles south of the takeoff point and that will take you 32 minutes (i.e. 60 x 80/150 = 32). During that 32 minutes the carrier will sail a further 17 miles.

Scenario 3 U.S. Planes have to fly to Midway Atoll
Midway Atoll is 255 miles @ 186° **To**.

Scenario 4 Japanese Carriers near Aleutian Islands Steam towards Midway
The distance between the Japanese northern fleet and the American fleet is (1,700 – 200 = 1500). To get within 200 miles to launch aircraft against the American fleet the Japanese would have to move 1,300 miles. However, the need to zig-zag would increase the distance traveled to 1,560 miles. At 35 knots this would take 44 hours and the overall course would be 225°.

Appendix 4
Constructing a Navy Pilot's Plotting Board
(for the navigation exercises in the final chapter)

In addition to the compass circle shown on the next page, you will need graph paper with squares, plus a tie tack and a China Marker (i.e. erasable grease pencil) as shown in Fig. 1. A fine point erasable felt pen can be used in place of the China marker, although they did not exist back in World War 2. Even if they had existed at the time, pilots would still have used the China marker, as they would not dry up at the wrong time.

If you do not have the capability to make a transparent copy of the compass circle, go to any office supply store like Office Max or Office Depot and have them make the copy. Have them enlarge it as much as possible but still fit on an 8.5" x 11" page as shown in Fig. 2. I recommend that you have them make several copies at the same time. You will note that the circle has mirror images of the numbers. This is deliberate. It means that the printed side will be face down when you use it, so that you will be drawing and erasing on the unprinted side.

Place the transparent compass circle on the graph paper so that the center is over a point where two major lines on the graph paper intersect. Then poke a hole through both with the tie tack. Remove the circle and draw a heavy line vertically and another horizontally through the rotation point on the graph. Mark 100, 200, 300 etc. on those lines. These are what I referred to as the distance lines in the instructions. It works best if the graph paper at attached to card stock or poster board to give it strength. I print the grid directly onto cardstock and then laminate it before attaching the circle.

You are now ready to do some dead reckoning navigation per the instructions in the final chapter of the book.

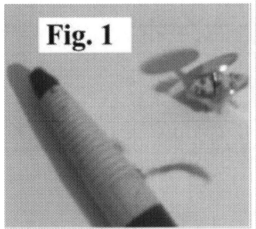
Fig. 1

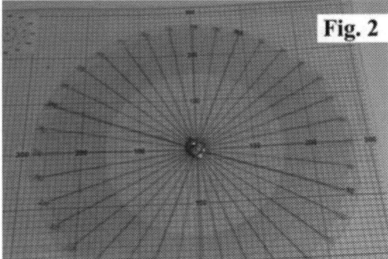
Fig. 2

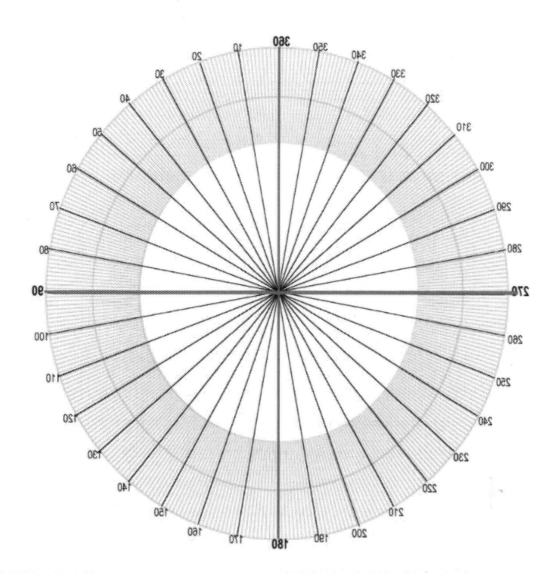